SpringerBriefs in Applied Sciences and Technology

SpringerBriefs present concise summaries of cutting-edge research and practical applications across a wide spectrum of fields. Featuring compact volumes of 50 to 125 pages, the series covers a range of content from professional to academic.

Typical publications can be:

- A timely report of state-of-the art methods
- An introduction to or a manual for the application of mathematical or computer techniques
- A bridge between new research results, as published in journal articles
- A snapshot of a hot or emerging topic
- An in-depth case study
- A presentation of core concepts that students must understand in order to make independent contributions

SpringerBriefs are characterized by fast, global electronic dissemination, standard publishing contracts, standardized manuscript preparation and formatting guidelines, and expedited production schedules.

On the one hand, **SpringerBriefs in Applied Sciences and Technology** are devoted to the publication of fundamentals and applications within the different classical engineering disciplines as well as in interdisciplinary fields that recently emerged between these areas. On the other hand, as the boundary separating fundamental research and applied technology is more and more dissolving, this series is particularly open to trans-disciplinary topics between fundamental science and engineering.

Indexed by EI-Compendex, SCOPUS and Springerlink.

More information about this series at http://www.springer.com/series/8884

Ali Salehabadi · Mardiana Idayu Ahmad ·
Norli Ismail · Norhashimah Morad ·
Morteza Enhessari

Energy, Society and the Environment

Solid-State Hydrogen Storage Materials

Springer

Ali Salehabadi
Environmental Technology Division
School of Industrial Technology
Universiti Sains Malaysia
Penang, Malaysia

Mardiana Idayu Ahmad
Environmental Technology Division
School of Industrial Technology
Universiti Sains Malaysia
Penang, Malaysia

Norli Ismail
Environmental Technology Division
School of Industrial Technology
Universiti Sains Malaysia
Penang, Malaysia

Norhashimah Morad
Environmental Technology Division
School of Industrial Technology
Universiti Sains Malaysia
Penang, Malaysia

Morteza Enhessari
Department of Chemistry, Naragh Branch
Islamic Azad University
Naragh, Iran

ISSN 2191-530X   ISSN 2191-5318 (electronic)
SpringerBriefs in Applied Sciences and Technology
ISBN 978-981-15-4905-2   ISBN 978-981-15-4906-9 (eBook)
https://doi.org/10.1007/978-981-15-4906-9

This Springer imprint is published by the registered company Springer Nature Singapore Pte Ltd.
The registered company address is: 152 Beach Road, #21-01/04 Gateway East, Singapore 189721, Singapore

Energy can neither be created nor destroyed; energy can only be transferred or changed from one form to another

—*Albert Einstein*

This book is dedicated to all students of IEK516 (Sustainable Energy Sources), M.Sc. (Environmental Science) Universiti Sains Malaysia

Preface

The contents of this book are meant to provide information, guide, and basic understanding of sustainable energy technology field focusing on hydrogen energy storage materials. It is written to inspire further research in this field. It can be used by senior undergraduate and graduate students, engineers, professionals, practitioners, scientists, researchers, planners, technologists, and employees in the area of engineering, technology, pure science, and applied sciences. It also can be used as a university reference book to serve as a graduate-level textbook to meet the growing demand for new courses in renewable and sustainable materials at technical and general universities. The book is divided into six chapters with each chapter provides good technical information.

Energy acts as the heart of the world and drives both natural and artificial mechanisms within it. Thus, it demands and causes significant impacts on our environment. Inappropriate energy use and production lead to main environmental problems, which are climate change and energy scarcity as well as a series of impacts. Hence, energy is an essential component of sustainable development. Due to this fact, the established sustainable development goals are in accordance with the energy system for holistic sustainable development actions. Chapter 1 reviews the energy, society, environment, and sustainable development. The relationship between energy and sustainable development are illustrated and discussed with the reference to the 2030 Sustainable Development Agenda. In short, energy and sustainable development must be studied and developed simultaneously in an integrated and comprehensive way to ensure the sustainability and wellness of our world.

Energy conversion is the key input for any proper consideration in energy production and consumption. The primary energy sources such as coal, natural gas, nuclear energy, petroleum, and renewable energy sources are used to generate secondary sources of energy like hydrogen. The secondary energy source is made from primary energy sources. Our need for energy to create order in the world stems from 1850, when Rudolf Clausius and William Thomson (Kelvin) stated the second law of thermodynamics. To order the disorderness and randomness of the natural tendency of matter and energy, a constant flow of quality energy through the system should be created. This orderness could be generated by nature and human society on Earth via their potency to structure and acquire energy. Chapter 2 covers these fundamental principles in detail.

Various types of storage technologies have been created so that the grid can achieve daily energy requirements. Ever since electricity was discovered, mankind has constantly searched for effective ways to store energy so that it can be used instantly when required. For the past century, technological advancement and shifting energy requirements have forced the energy storage industry to adapt and evolve. Chapter 3 discusses an overview and types of energy storage systems.

Hydrogen is an ideal candidate to fuel as "future energy needs". Hydrogen is a light (Mw = 2.016 gr mol^{-1}), abundant, and nonpolluting gas. Hydrogen as a fuel can be a promising alternative to fossil fuels, i.e., it enables energy security and takes care of climate change issues. Hydrogen has a low density of around 0.0899 kg m^{-3} at normal temperature, and pressure ($\sim$7% of the density of air), which is the main challenge in its real applications. It means, for example, 1 kg of hydrogen requires an extremely high volume of around 11 m^3. In order to solve this limitation of hydrogen, solid-state hydrogen storage materials are used to store hydrogen efficiently and effectively. In Chap. 4, an attempt has been developed to provide a comprehensive overview of the recent advances in hydrogen storage materials in terms of capacity, content, efficiency, and mechanism of storage.

In Chap. 5, a brief description of the requirements of a hydrogen storage system is given. The weak interaction of hydrogen within pores (sites) needs to be understood in order to design and develop porous materials for hydrogen sorption. The measurements are based on the amount of hydrogen adsorbed as a function of pressure, temperature, the enthalpies of adsorption, and the adsorption/desorption characteristics.

The term "energetic materials" are a class of materials that can release stored molecular chemical energy *via* external stimulations or internal modifications. We aim to take advantage of these opportunities by bringing some logical ideas onto/into the surface of hydrogen storage materials. In addition, hydrogen energy storage systems provide multiple opportunities to enhance the flexibility and improve the economics of energy supply systems in the electric grid, gas pipeline systems, and transportation fuels; therefore, it is critical to boost hydrogen storage performance of the materials. The high mobility of the hydrogen and their variable compositions can be enhanced by improving the properties of the host media. In Chap. 6, the most important factors, which can affect hydrogen storage performance of the solid-state materials are discussed.

In the process of preparing and writing this book, the support provided by the individuals and institutions is noteworthy. In this context, we would like to express our appreciation to Springer Publishing Editor Dr. Loyola D'Silva, Project Coordinator Mr. Ashok Arumairaj, and all the editorial team of Springer Nature for their contribution in any kind of forms in bringing the book to fruition. Our thanks go to the School of Industrial Technology, Universiti Sains Malaysia and the Department of Chemistry, Islamic Azad University, Naragh Branch, Naragh, Iran for facilitating the process of gathering material and information for publishing this book. In addition, we also appreciate all reviewers for their time reviewing the content of this book. We also thank our families for their patience and support during the preparation of this book. Our special thanks go to Ms. Ang Jia Hui for helping us gathering materials for some chapters of this book. This book would not be possible without their kind support from many aspects during the process.

The work involved in this book is part of the outcomes of funded research projects, thus we would like to take this opportunity to convey our appreciation to the sponsors for the financial and technical supports. Our thanks go to USM Research University Grant and TRGS and FRGS Grant Ministry of Education (MOE) Malaysia. Any opinions, findings, and conclusions or recommendations expressed in this material are those of the authors and do not necessarily reflect the views of the MOE and USM. Last but not least, it is hoped that this book would serve as a valuable guide to academics, researchers, professionals, and students working in this field.

Penang, Malaysia   Ali Salehabadi
Penang, Malaysia   Mardiana Idayu Ahmad
Penang, Malaysia   Norli Ismail
Penang, Malaysia   Norhashimah Morad
Naragh, Iran   Morteza Enhessari
March 2020

Contents

About the Authors

Ali Salehabadi obtained his Ph.D. in Polymer Chemistry from School of Chemical Sciences, Universiti Sains Malaysia in 2014. He is currently a postdoctoral fellow in the Environmental Technology Division, School of Industrial Technology, Universiti Sains Malaysia. He is a chemist with the background of polymer chemistry and solid-state energetic materials (Nanomaterials, MOFs, Polymers), with application in storage and solar systems. His academic and research experiences provided him with a strong background in multiple disciplines in chemistry, advanced materials, and environments. He has published more than 40 papers in top-tier journals including three US-patents, and chapter books. e-mail: alisalehabadi@usm.my

Mardiana Idayu Ahmad obtained her Ph.D. in Engineering Science: Sustainable Energy Technologies at the Department of Architecture and Built Environment, Faculty of Engineering, University of Nottingham, United Kingdom in 2011. She is currently an Associate Professor in the Environmental Technology Division, School of Industrial Technology, Universiti Sains Malaysia. Her research spans in the breadth of sustainable energy technologies and environmental management. She has always been passionate about continuing her research in a way to bridge these two fields. Her research work leads to the production of over 100 publications nationally and internationally, including journal papers, research books, popular academic books, book chapters, conference proceedings, and other publications. e-mail: mardianaidayu@usm.my

Norli Ismail is currently the Head of Department and professor in the School of Industrial Technology, University Sains Malaysia. She obtained her Ph.D. in Environmental Technology from Universiti Sains Malaysia. Her research interests are biogas production, water, wastewater, solid waste treatment technology, and management. e-mail: norlii@usm.my

Norhashimah Morad earned her Bachelor of Science in Chemical Engineering from University of Missouri-Columbia, USA, in 1985. She joined the School of Industrial Technology in University Sains Malaysia (USM), Malaysia as a tutor in 1988 and pursued her Ph.D. in Control Engineering at the University of Sheffield, UK under the Commonwealth Scholarship. She is currently a Professor in Environmental Technology Division, USM, Malaysia. Her research interests are life cycle assessment (LCA), intelligent systems in manufacturing, optimization using genetic algorithms, phytoremediation, and new methods and materials in biological and chemical wastewater treatment. e-mail: nhashima@usm.my

Morteza Enhessari earned his Ph.D. in inorganic chemistry from the Science and Research branch, Islamic Azad University, Tehran, Iran, in 2008. He is currently an Associate Professor of Chemistry at Naragh Branch of Islamic Azad University. His research interests are pioneered in the synthesis and characterization of Perovskite Nanopowders. He was also awarded "Highly commended paper 2014 (Emerald Group Publishing Limited)" for his work on $Cr_{1.3}Fe_{0.7}O_3$ nanopigments. Enhessari has over 50-refereed publications and three granted US-patents. e-mail: enhessari@iau-naragh.ac.ir

Chapter 1
Overview of Energy, Society, and Environment Towards Sustainable and Development

Abstract Energy acts as the heart of the world and drives both natural and artificial mechanisms within it. Thus, it demands and causes significant impacts on our environment. Inappropriate energy use and production lead to main environmental problems, which are climate change and energy scarcity as well as a series of impacts. Hence, energy is an essential component of sustainable development. Due to this fact, the established sustainable development goals are in accordance with the energy system for holistic sustainable development actions. Therefore, this chapter reviews the energy and sustainable development. The relationship between energy and sustainable development was illustrated and discussed with the reference to the 2030 Sustainable Development Agenda. In short, energy and sustainable development must be studied and developed simultaneously in an integrated and comprehensive way to ensure the sustainability and wellness of our world.

1.1 Introduction

Energy is the primary driver of the natural and anthropogenic processes of the Earth. According to the First Law of Thermodynamics, energy cannot be created or destroyed, however, it can be transferred and converted. Energy can be transformed into other forms to power specific works. The most in-demand forms of primary energy are oil and coal, at 32% and 26% of the total energy, respectively (Enerdata 2020). The primary energy can be used directly after it was extracted or converted into secondary energy, such as electricity and petroleum. These energies commonly power the work and dissipate the heat. Positively, heat energy can be collected and regenerated via systems that are distinguished by their energy efficiencies (Arthur et al. 2018). However, the mismatch of heat and system causes insufficient energy in its useful form and hence, global temperature rises.

Along with rapid population growth, development, and urbanization, the increasing use of energy leads to energy scarcity, which causes significant impacts on energy production and utilization. Concurrent with the increasing energy demand, carbon emissions also increased by 2.0% in 2018 (BP 2019). Global energy demand and global primary energy increased by 2.3% and 2.8%, respectively in 2018, which are record increments in the last ten years. The total global energy consumption is up to 13,820 Mtoe in 2018, led by Asia at 5859 Mtoe (Enerdata 2020). Among the global sectors' energy demands, the industry and building sectors contributed the highest percentages, which are 43% and 29%, respectively, in 2017. The industry and building sectors are predicted to reach 42% and 32%, respectively, in 2040 under the evolving transition scenario, while 39% and 33%, respectively, in 2040 under the rapid transition scenario (BP p.l.c. 2019). Thus, policies should focus on the industry and building sectors.

Inappropriate energy production and use cause global environmental problems, such as climate change, pollution, health problems, due to the emission of waste products and natural resource depletion (IEA 2019a). These environmental problems lead to chain effects as follows: health problems, inclination of learnability, decrease talents to contribute to society, and global inequality (Perera 2018; UNEP 2020).

Energy is the main aspect of sustainable development as reported in the World Energy Outlook report and of the sustainable development scenario stated in the Sustainable Development Agenda 2030. The three pillars of sustainable development are environmental, economic, and social. It is the main factor of climate change, environmental pollution, as well as the development of facilities and infrastructures for industries, services, and institutions (Omer 2014). Therefore, energy and sustainable development are vital to be studied and developed simultaneously as a whole to achieve sustainable development, ensuring the fulfilment of the present while not compromising the need for the future (IEA 2019a; Prag 2018).

Population growth and spreading of environmental problems urge sustainable development in accordance with research and development (R&D), as well as the production and use of sustainable, renewable, and energy-efficient processes. Renewable energy occupies up to 15% of the total global energy used in 2017 and was predicted to increase to 26% and 44% in 2040 under the evolving transition scenario and rapid transition scenario, respectively (BP p.l.c. 2019). The current established policies, namely the Paris Agreement and Sustainable Development Goal, work based on the energy system transformation for sustainable, renewable, and efficient energy (McCollum et al. 2018). The Sustainable Development Scenario (SDS), stated in the Sustainable Development Agenda 2030, presented the approaches to meet the SDGs, especially the energy-related goals, via the swift and extensive transformation of the energy system worldwide. The SDS aims for the mean global temperature rise to be below 1.65 °C by reducing greenhouse gases (GHGs) emissions especially from energy production and use. The global carbon dioxide emission was predicted to drop from 33 billion tonnes to 10 billion tonnes from 2018 to 2050, achieving net-zero emissions by 2070 (IEA 2019b). Among the 17 Sustainable Development Goals (SDGs), there are three main energy-related goals, namely the 7th goal: affordable and clean energy, the 3rd goal: good health and well-being, and the 13th goal: climate action. The remaining 14 goals are closely energy-related and slightly energy-related goals. All of the ways of implementations are in accordance to the Paris Agreement, maintaining the mean global temperature below 2 °C and further reach the limit below 1.5 °C. Therefore, this chapter reviews the relationships between energy and sustainable development.

1.2 Energy and Sustainable Development

1.2.1 The 2030 Sustainable Development Agenda

Currently, the main efforts towards sustainable development are the 2030 Agenda for sustainable development, which was taken on by Heads of State and Government at the United Nation summit in 2015. It is a 15-year global framework made up of 17 Sustainable Development Goals (SDGs), 169 targets, and 230 indicators. The main objective of this agenda is to eliminate poverty and achieve sustainable development by 2030 extensively throughout the globe, including all participants. It was designed to be applicable for both developing and developed countries, working in a holistic way. The core principles of the 2030 agenda are universality, leaving no one behind, interconnectedness and indivisibility, inclusiveness, and multi-stakeholder partnerships. It was implemented based on five dimensions, namely people, planet, prosperity, peace, and partnership.

It combines the three pillars of sustainable development, namely social, economic, and environmental, as well as peace, governance, and justice components. 2030 Agenda for sustainable development is divided into four sections, namely political declaration, 17 Sustainable Development Goals and 169 targets, ways of implementation, and a framework of follow up and review to evaluate the actions related to the agenda (European Commission 2017; United Nations 2015, 2018). The sustainable development goals in the 2030 agenda are directly or indirectly related to energy issues. The direct relation of the energy and the sustainable development goal is observed in the 7th Goal: Affordable, Clean Energy, 3rd Goal: Good Health and Well-being, 4th Goal: Quality Education, and 13th Goal: Climate Action. The indirect relations of energy and the sustainable development goal can be seen in the remaining goals as shown in Fig. 1.1.

1.2.2 Direct Energy-Related Sustainable Goals

SDG 7 Affordable and clean energy aims to expand access to affordable, reliable, sustainable, and modern energy for all, including electricity and fuels. This was indicated by the population with access to electricity as well as clean fuels and technology. Thus, the development of sustainable and efficient energy technologies should be done in both developed and developing countries to provide a continuous and accessible electricity supply. Developed countries have a large energy generation capacity but it was compensated by their high-energy consumption. To date, there is 11% of the developing countries do not access to electricity (United Nations 2019b). The sustainable energy was developed based on several requirements, namely safe and clean, affordable cost, high accessibility, and renewable, as well as high efficiency (Columbia Center on Sustainable Investment 2019; Dincer and Rosen 1999). For example, the solar, wind, and hydro energy are renewable,

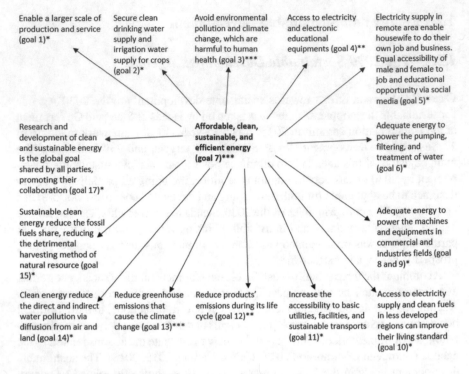

Fig. 1.1 Relationship of energy to the related sustainable development goals

have low impacts, affordable cost, high availability, and efficient, if they are applied based on the environmental conditions and the target users' needs. This goal also targets to significantly increase the percentage of renewable energy out of the total final energy consumption. Global enhancement of energy efficiency is targeted to be increased by twofold, in terms of primary energy and Gross Domestic Product (GDP). In addition, the international collaborations in the research and development (R&D) of sustainable, renewable, and clean energy should be implemented to ensure the international financial flows to compensate the financial limitation of the developing countries. Along with this, local and international investments in the facilities, infrastructures, and technologies were targeted to ensure the availability of the modern and sustainable energy service for all societies in developing countries and least developed countries and states. This facilitates their energy and sustainable development efforts (United Nations 2019a). To date, the actions taken by the member countries show good progress. In 2019, the renewable energy occupied 17.5% of the global total energy use, while the global primary energy intensity was increased to 5.1%. Moreover, it was reported to achieve up to 18.6 billion international flow. In 2019, the global accessibility to electricity was increased to 87%, while the global accessibility to the clean fuels and technologies was increased to 61%. However, there is still a gap between the current achievement and the targeted achievement of this goal (United Nations 2019a).

SDG 3 good health and well-being can be improved by using clean and safe energy such as solar energy, hydro energy, wind energy, nuclear energy, and artificial kinetic energy. These energy technologies prevent environmental pollutions that have enormous harmful effects on human and other living organisms. For instance, transportation and industrials activities record lower GHGs emissions along with the increased use of clean and sustainable energy. Moreover, the electrification and clean fuel technologies prevent indoor air pollution caused by the smoke from conventional cooking fuels (IEA 2019a). The clean energy also reduces the explosion and fire risk caused by the combustion of fossil fuels. By 2017, the results of the actions taken under goal 3 are still insufficient to hit the targets. There are up to 7 million annual death caused by fine particles air pollutions, especially indoor air pollution (UNDP 2019).

One of the targets in SDG 4: Quality education is that the education facilities should be safe, nonviolent, inclusive, and suitable for effective learning. This was indicated by its access to multiple facilities and infrastructures, including electricity supply, which is one of the main forms of energy. Most of the developing countries are still facing inadequate basic facilities and infrastructure for productive learning environments. For example, it was reported that there is less than 50% of the schools that access to electricity, drinking water, computers, and internet in Sub-Saharan African (United Nations 2019a).

In the SDG 13: climate action, a series of actions were being taken to mitigate and adapt the climate change via regulation of greenhouse emissions, as well as the development of clean and renewable energy (European Commission 2017). The reduction of air pollution reduces global warming besides than the extreme weather and climate. In 2016, the Paris Agreement was established and signed to ensure the rise in global mean temperature remains below 2 °C by reducing greenhouse gas emissions, as well as to adapt to the climate change effects. In 2017, up to 143 parties ratified the Paris Agreement, and 137 parties submitted their first nationally verified contributions to the United Nations Framework Convention on Climate Change (UNFCCC). The number of participating countries was increased to 168 and 185 parties in 2018 and 2019, respectively. Up to 28 countries had accessed the Green Climate Fund grant to support implement their adaptation plans in 2019 (United Nations 2020).

1.2.3 Indirect Energy-Related Sustainable Goals

Moreover, energy development can indirectly facilitate in achieving all sustainable development goals. For instance, in the SDG 12: responsible consumption and production, energy as one of the main natural resources, should be managed and used efficiently and wisely. The development of high-efficiency renewable energy technologies provides the maximum, safe, and secure energy output at the lowest cost.

For example, solar energy harvesting in regions with consistent long solar hours and intensive solar penetration provide maximum energy production. Waste production of the energy source during its life cycle from extraction, production, and transportation to its end users is also mentioned in this goal. This can be achieved by using selective clean energy technologies with the least waste product such as geothermal energy, hydro energy, wind energy, and artificial kinetic energy. Solar energy and nuclear energy are less recommended energy technologies to achieve this goal since they produce waste and high-temperature water. These technologies provide affordable and secure electricity to commercial, domestic, and industrial use. This contributes to the reduction of fossil fuel use, for example, the replacement of conventional cooking fuel by the electric induction cooker, as well as the introduction of the hybrid engine car. Another example is that energy development can ensure the water treatment and pumping to supply clean and secure water to all, promoting the sustainable cities and industrialization (United Nations 2015). This promotes economic growth, social well-being, and equality, in addition to terrestrial and aquatic biodiversity health.

1.2.4 Implications

To date, many actions are taken under energy development in accordance with sustainable development goals across the world. For instance, in 2019, the installation of solar panels systems was done in the Lelepa village in Vanuatu. It was made up of three cubes with 200 W of power storage each. The power cubes can compensate each other during the failure of any cube. This action fulfills the villagers' electricity needs, enable longer storage of fishery products, pumping of clean water to the local buildings, commercialized sewing and handicraft works by the women, printing and computer jobs in educational institutions, as well as the cold storage for the vaccines and medicines (UN Development Programme 2019b). This simultaneously facilitates in reaching the goals of SDG 1, SDG 2, SDG 3, SDG 4, SDG 5, SDG 6, SDG 7, SDG 8, SDG 9, SDG 10, and SDG 13.

By 2019, wise energy usage in Bhutan only produces relatively low greenhouse emissions from the transportation and industrial sector. The reforestation program held in Bhutan acts as a carbon sink in which forest functions to absorb all the greenhouse emissions. Thus, Bhutan is considered to have zero emissions and reached the target of the Paris Agreement (UN Development Programme 2019a). This simultaneously facilitates in reaching the SDG 3, SDG 13, SDG 14, and SDG 15.

At the same time, Chile also shows great efforts in reducing global carbon emissions via the development of clean sustainable energy. It was reported to achieve the target to have only less than 1% of the global carbon emissions. By 2019, up to 200 electric buses were used in Chile. Chile even aims for 80% of electric public transport by 2022. Simultaneously, 28 coal power plants were planned to be abolished by 2040. This shows Chile's intention to convert its energy dependency towards clean sustainable energy, leading to its carbon neutrality goal by 2050 (UN Development Programme 2019a). This simultaneously facilitates in reaching SDG 3, SDG 13, SDG 14, and SDG 15.

1.3 Summary

In short, energy initiates all the processes and cycles, so it is the root factor of the impacts as well as the sustainability of the world. Therefore, R&D as well as the appropriate production and use of energy, are crucial to the achievement of all SDGs. Without a doubt, the relationship between energy and SDGs is compulsory to be studied to produce sustainable development actions, which is done in accordance with the transformation of the energy system.

Acknowledgements Universiti Sains Malaysia research grant (1001/PTEKIND/8014124), and the authors would like to thank Ms. Ang Jia Hui.

References

S. Arthur, C. Doyle, A. Study, Thermodynamics for Beginners—Chapter 1 THE FIRST LAW If you were asked to prove that two and two made four, you might (2018)

BP, Full report—BP Statistical Review of World Energy 2019 (2019)

BP p.l.c., BP Energy Outlook 2019. BP Energy Outlook 2019 (2019)

Columbia Center on Sustainable Investment, The Energy Sector and the Sustainable Development Goals (2019)

I. Dincer, M.A. Rosen, Energy, environment and sustainable development. Appl. Energy **64**, 427–440 (1999)

Enerdata, World Energy Consumption Statistics (2020). Retrieved from: https://yearbook.enerdata.net/total-energy/world-consumption-statistics.html. 14 Jan 20

European Commission, The 2030 Agenda for Sustainable Development and SDGs—Environment—European Commission (2017). Retrieved from: https://ec.europa.eu/environment/sustainable-development/SDGs/index_en.htm. 2 Jan 20

IEA, Sustainable Development Scenario—World Energy Model. World Energy Outlook 2019 (2019a). Retrieved from: https://www.iea.org/reports/world-energy-model/sustainable-development-scenario

IEA, World Energy Outlook 2019 (2019b). Paris. Retrieved from: https://www.iea.org/reports/world-energy-outlook-2019

D.L. McCollum, W. Zhou, C. De Bertram, H.S. Boer, V. Bosetti, S. Busch, J. Després, L. Drouet, J. Emmerling, M. Fay, O. Fricko, S. Fujimori, M. Gidden, M. Harmsen, D. Huppmann, G. Iyer, V. Krey, E. Kriegler, C. Nicolas, S. Pachauri, S. Parkinson, M. Poblete-Cazenave, P. Rafaj, N. Rao, J. Rozenberg, A. Schmitz, W. Van Schoepp, D. Vuuren, K. Riahi, Energy investment needs for fulfilling the Paris Agreement and achieving the Sustainable Development Goals. Nat. Energy **3**(7), 589–599 (2018). https://doi.org/10.1038/s41560-018-0179-z

A.M. Omer, Sustainable development and environmentally friendly energy systems in sudan. Adv. Environ. Res. **36**(1), 51–94 (2014)

F. Perera, Pollution from fossil-fuel combustion is the leading environmental threat to global pedi-
    atric health and equity: solutions exist. Int. J. Environ. Res. Public Health **15**(1) (2018). http://
    doi.org/10.3390/ijerph15010016
A. Prag, The IEA Sustainable Development Scenario (2018)
UN Development Programme, Five Plans for Carbon Neutrality (2019a). Retrieved from: https://
    medium.com/@UNDP/five-plans-for-carbon-neutrality-f61391ce2228. 7 Jan 2020
UN Development Programme, Powering Change in Vanuatu (2019b). Retrieved from: https://
    medium.com/@UNDP/powering-change-in-vanuatu-48f22e38a762. 6 Jan 2020
UNDP, Goal 3: Good Health and Well-Being (2019). Retrieved from: https://www.undp.org/content/
    undp/en/home/sustainable-development-goals/goal-3-good-health-and-well-being.html. 15 Jan
    2020
UNEP, Results of the UNEP foresight process on emerging environmental issues (2020). Retrieved
    from: https://www.unenvironment.org/resources/report/21-issues-21st-century-results-unep-
    foresight-process-emerging-environmental. 10 Jan 2020
United Nations, The 2030 Agenda for Sustainable Development (2015). Retrieved from: www.
    unssc.org
United Nations, Transforming Our World: The 2030 Agenda for Sustainable Development. A New
    Era in Global Health (2018)
United Nations, Goal 4 Sustainable Development Knowledge Platform (2019a). Retrieved from:
    https://sustainabledevelopment.un.org/sdg4. 5 Jan 2020
United Nations, Goal 7 Sustainable Development Knowledge Platform: Progress of Goal 7 in 2019
    (2019b). Retrieved from: https://sustainabledevelopment.un.org/sdg7. 8 Jan 2020
United Nations, Goal 13 Sustainable Development Knowledge Platform: Progress of Goal 13 in
    2019 (2020). Retrieved from: https://sustainabledevelopment.un.org/sdg13. 4 Jan 2020

Chapter 2
Overview of Energy

Rudolf Julius Emanuel Clausius (1822–1888), a German physicist who is considered one of the central founders of law of thermodynamics

William Thomson, 1st Baron Kelvin (1824–1907), an Irish–Scottish mathematical physicist who extended the Carnot Clapeyron theory

Albert Einstein (1879–1955), a German theoretical physicist, who developed the theory of relativity, and the mass–energy equivalence formula ($E = mc^2$)

Abstract Energy conversion is the key input for any proper consideration in energy production and consumption. The primary energy sources such as coal, natural gas, nuclear energy, petroleum, and renewable energy sources are used to generate secondary sources of energy like hydrogen. The secondary energy source made from primary energy sources. Our need for energy to create order in the world stems from 1850, when Rudolf Clausius and William Thomson (Kelvin) stated the second law of thermodynamics. To order the disorderness and randomness of the natural tendency of matter and energy, a constant flow of quality energy through the system should be created. This orderness could be generated by nature and human society on Earth via their potency to structure and acquire energy. This chapter will cover these fundamental principles in detail.

A. Salehabadi et al., *Energy, Society and the Environment*,
SpringerBriefs in Applied Sciences and Technology,
https://doi.org/10.1007/978-981-15-4906-9_2

2.1 Introduction

In today's society, we are facing the energy dilemma, mainly in three components such as oil, environment, and expanding global demand. The majority of world energy still comes from petroleum, mostly from the Middle East, which is not unlimited, and we are so close to the peak looms of oil production (Kuppusamy et al. 2020b). But then, when people got oil from the ocean floor or on land, other chemicals and toxic substances come up, such as mercury, lead and arsenic, and organic compounds (Kuppusamy et al. 2020a). These materials are often released back into the ocean, remain in the soil, or burn into the atmosphere, which directly affects the environment. These toxic substances cover the ocean and are enhanced by solar UV radiation and finally cause a great negative impact on food productivity. Global population is forecast to rise gradually by the near future. This population affects the current growth in energy use. Civilization and the developing world need more and more energy to be produced (Mullan and Haqq-Misra 2019). The right solution to these dilemma face into three entangling factors; the progress toward alternatives to petroleum sources is slow, the social norms and uncertainty change is difficult, and the acting time has already lapsed. Hence, it is urgently required that these major global deficiencies concerning energy and the environment face together, and the simplest solution to both problems is the replacement of petroleum and petroleum products with an environmentally friendly fuel, like secondary energy sources (Davidson 2019). Among several advantages of renewable energy sources as compared to nonrenewable energy sources, however, primary-renewable energy sources are still suffering from the various drawbacks like higher upfront cost, intermittency, storage capabilities, and geographic limitations. Figure 2.1 simply represents the relation between energy and human life.

What is energy? Energy is the capacity to do work and a core element of the interaction between humanity and nature. Nature and society are the structured and thorough result of all the main energy sources, flows, storages, and conversions that molded the development of life on Earth (Fig. 2.2). Energy is a quantitative

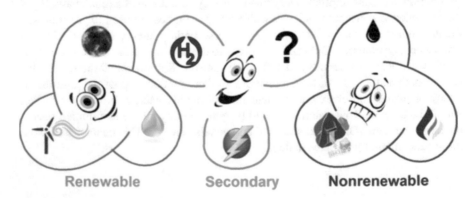

Fig. 2.1 Energy and human civilization

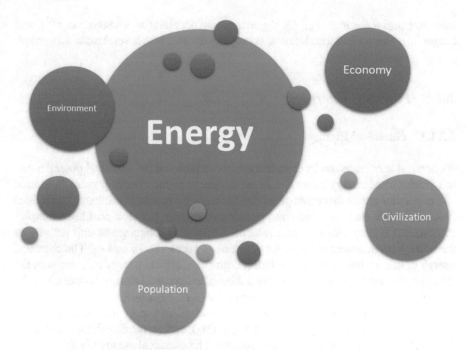

Fig. 2.2 Correlation between energy, nature, and society

characteristic that must be shifted to an object in order for work to be carried out on it. Energy can be conserved, as per the Law of Conservation of Energy, which states that energy can change forms but cannot be created or destroyed. Einstein's equation supports the relation between energy (E) and mass of the matter (m) as Eq. 2.1

$$E = mc^2, \qquad (2.1)$$

where mass and energy are directly proportionate to each other.

2.2 Forms of Energy

Energy exists in several forms but classified into two main and basic forms; potential and kinetic. Potential energy is stored energy and the energy of position. This is the force required to raise an object, against the force of gravity involve two main factors; forcing time and distance. The potential energy (E_p) can be formulated as Eq. 2.2

$$E_p = \text{force} \times \text{distance} = \text{weight} \times \text{height} = mgh, \qquad (2.2)$$

here m, g and h are mass (g), local gravitational accelerating (9.80665 m s^{-2}), and height (m). The potential and kinetic energy are divided into several main subgroups.

2.2.1 Potential Energy

2.2.1.1 Chemical Energy

A chemical substance has its chemical energy in terms of the chemical potential and has its chemical exergy as well. Chemical energy is the energy present in atomic and molecular bonds that often generates heat as a by-product (exothermic reaction) (Schmal 2014). Batteries, biomass, petroleum, natural gas, and coal are examples of chemical energy. For instance, assume a chemical substance at unit activity in the normal environment at initial temperature and pressure T_1 and P_1. The chemical exergy of this system is in close relation with the chemical exergy references in the atmosphere, seawater, and lithosphere solids. The chemical exergy is defined as the maximum work, which can be supplied from chemical equilibrium with the reference environment at constant temperature and pressure. The exergy reference species in the atmospheric air are oxygen (O_2), nitrogen (N_2), carbon dioxide (CO_2), and water vapor (H_2O) at their respective concentrations. The chemical exergy ($\acute{\epsilon}$) of a pure O_2 at the unit activity can be calculated from Eq. 2.3

$$\acute{\epsilon} = -RT_1 \ln x, \tag{2.3}$$

where R is the gas constant and x is the molar fraction of O_2.

In chemical energy, there are two important reaction mechanism called exoergic and endoergic reactions; the latter is a reaction that releases energy when chemical bonds form, for example, exothermic nuclear reaction, and the former is one that requires an input of energy to take place among an endothermic reaction, for example, photosynthesis reaction. In photosynthesis, the energy from the sun is served by the trees to break down the CO_2 and H_2O bonds. The change in free energy in exoergic is negative ($\Delta G < 0$), where the energy is released, while endoergic is positive ($\Delta G > 0$), and the energy is absorbed (Fig. 2.3).

2.2.1.2 Nuclear Energy

Nuclear energy is the energy present in an atom's nucleus that binds it together. When atomic nuclei are fused or separated, massive amounts of energy are released. Atoms are made up of three particles: protons (positive electrical charge), neutrons (no electrical charge), and electrons (negative electrical charge). An atom has a core (nucleus) containing protons and neutrons. This core is surrounded by electrons. All nuclei are bonded to each other, and energy is present in the bonds that keep the nucleus together. Nuclear energy can be generated through the breaking of these

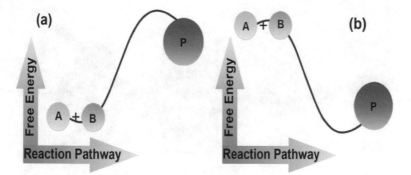

Fig. 2.3 a Endoergic and **b** exoergic reaction mechanism; in these two reaction mechanisms "A" and "B" are starting materials (reactants) and "P" is the products. Endergonic reaction is a nonspontaneous reaction and requires energy to drive, while vice versa in exergonic reaction

bonds by nuclear fission, and this energy can be utilized to produce electricity (Şahin and Mehmet Şahin 2018).

Electrical and gravitational forces are present in the nucleus. Electrical forces attempt to push protons apart while gravitational forces bring the nucleons together. There are powerful forces that hold the nucleus together. The energy requires for breaking nucleus, and separating into proton and neutrons are called nuclear binding energy. Einstein's equation clearly explains this energy (Eq. 2.4)

$$E = m \times c^2,\qquad(2.4)$$

where E, m, and c are energy (kJ), mass (kg), and speed of light (2.998×10^8 m^2 s^{-1}), respectively.

The most commonly used nuclear fuel for nuclear fission in nuclear plants is uranium. Though it is relatively common and can be found in rocks anywhere in the world, it is a nonrenewable energy source. In a typical nuclear power plant, U-235 is used as an active fuel because its atoms are easily split apart. After the uranium is mined, the U-235 will be extracted and processed before it can be used as a fuel.

All elements with more than 83 protons are naturally radioactive, having alpha, beta, and gamma radiations. The nuclear power plant is work among the following two processes:

- **Nuclear fission**—The breaking of a big atomic nucleus into many smaller nuclei. In a nuclear reactor, a neutron will mostly be absorbed into a nucleus (usually of uranium-235), causing the nucleus to transform into the extremely active uranium-236. This large nucleus will break into two big fragments known as "daughter nuclei". In addition to these "daughter" products, two or three neutrons remain intact. These neutrons can start a new fission reaction known as a chain reaction (Şahin and Wu 2018). Figure 2.4 shows the process of nuclear fission.
- **Nuclear fusion**—The process of nuclei combination, irrespective of nuclear fission. In the fusion process, two small nuclei join together to form a heavy

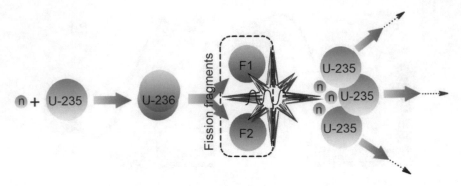

Fig. 2.4 Process of nuclear fission

nucleus (Wu and Şahin 2018). Nuclear forces are short in distance and have to act against the electrostatic forces produced when positively-charged nuclei repel other positively-charged nuclei. This is why a successful nuclear fusion occurs at a very high density and temperature (Fig. 2.5). Fusion reactions commonly occur in stars, where two hydrogen nuclei combine under high pressure and temperature to create a helium isotope nucleus. In the sun, the simplest fusion happens when four hydrogen nuclei become one helium nuclei (Eq. 2.5)

$$4_1^1H \rightarrow {}_2^4He + 2\beta^+ + energy. \tag{2.5}$$

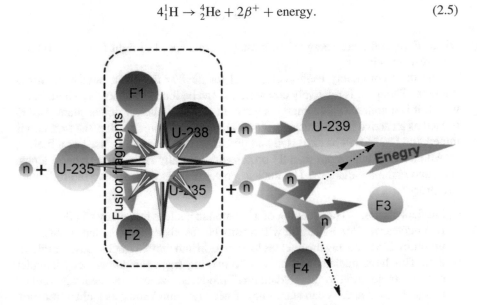

Fig. 2.5 Process of nuclear fusion

The combined mass of four hydrogen nuclei is 6.693×10^{-27} kg, while the mass of one helium nucleus is 6.645×10^{-27} kg. The differentiation between these two masses is equal to 0.048×10^{-27} kg. This missing mass can be converted to energy, which radiates away.

From both, nuclear fission and fusion, a high amount of energy can be released; however, the amount of released energy from nuclear fusion is much higher than that of nuclear fission.

2.2.1.3 Gravitational Energy

Energy is also stored in an object when it has a higher position compared to a lower position (Fig. 2.6). Progress in gravitational energy studies started in the 1950s. Gravitational energy is the energy stored in an object's height (York 1980). The higher and heavier the object, the more gravitational energy stored in it. As the name suggests, gravitational energy is borne from gravity. Gravity is a force that pulls an object towards another object. For example, Earth's gravity is the reason objects fall and it is also the force keeping people on the ground. The gravitational potential energy ($E_{GPE.}$) of an object of mass m at height h on Earth can be calculated by Eq. 2.6

$$E_{PE.} = mgh. \tag{2.6}$$

Physically, the force applied to the object is an external force. This force can do positive work, which can further increase the gravitational potential energy. In gravitational potential energy calculation, there is an assumed reference level with the potential energy equal to zero. This level is, in fact, the Earth's surface, while

Fig. 2.6 Gravitational potential energy, the work done to lift the weight is stored in the mass-Earth system as gravitational potential energy

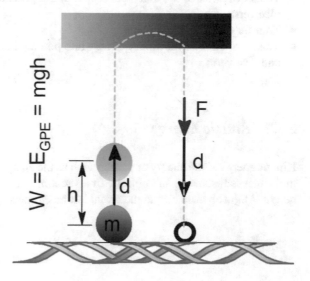

it is arbitrary. The difference in gravitational potential energy is related to the work done. Gravitational potential energy may be converted to other forms of energy. If we release the mass, gravitational force will do an amount of work equal to *mgh* (Livadiotis 2017).

2.2.1.4 Mechanical Energy

Humans and animals need energy to move, which they obtain when chemical energy is converted into mechanical energy. For living things, movement is a critical factor in terms of energy conservation. The source of mechanical energy is anything acting (force/movement) on the system to change mechanical energy (Latash and Zatsiorsky 2016).

Mechanical energy is a functional combination of both potential and kinetic energy. It is defined as the energy possessed by an object due to its motion, as well as the energy stored in objects from tension. Compressed springs and stretched rubber bands are examples of stored mechanical energy. Mechanical energy is the product of force (F) and distance (Δl). When stored energy is released, the body moves in the direction of the force.

As mentioned before, mechanical energy is both kinetic and potential energies in one system. This energy, like many other forms of energy, should convert into other forms of energy, for example, steam engines convert mechanical energy into heat energy, or turbines convert the steam (heat energy) into mechanical energy. Hydroelectric power plants are one of the most important examples in mechanical energy conversion, where electrical energy is formed from the mechanical energy of the stored water.

Let us count the mechanical energy and its mechanism of a typical motion:

- Transformation of the kinetic energy into the gravitational potential energy and vice versa.
- Transfer of mechanical energy.
- Transformation of the kinetic energy into potential energy (deformation process) and vice versa.

2.2.2 Kinetic Energy

Kinetic energy is the energy of motion such as the energy of waves, electrons, atoms, molecules, substances, and objects. In a vacuum, when the mass (m) drops from a height (h), just before hitting the ground, it has a speed (U) of Eq. 2.7

$$U = (2gh)^{1/2} \tag{2.7}$$

This is the point that the potential energy converts into the kinetic energy as Eq. 2.8

$$E_p = mgh = \frac{1}{2} mU^2 \qquad (2.8)$$

Five main subgroups of kinetic energy are:

Radiant energy: Electromagnetic energy that moves in the form of transverse waves. Examples of radiant energy include X-rays, visible light, gamma rays, and radio waves. Light is considered radiant energy; this makes sunshine, a primary source of fuel and warmth for life on Earth, a form of radiant energy too.

Thermal energy (heat): The energy generated from atomic and molecular movement in an object. The speed of these particles increases the object's heat. Geothermal energy is the thermal energy present in the Earth.

Motion energy: The energy present when objects move. The greater an object's speed, the more energy it generates. Energy is required to make an object move and this energy is released when the object slows down. Wind is a popular example of motion energy, as is a car crash, where a vehicle stops abruptly and releases all of its motion energy at one go.

Sound: The energy that moves through objects lengthwise, either by rarefaction or compression waves. Sound is generated when a force causes an object to vibrate, and energy travels through the object in the form of a wave. Generally, sound energy is smaller compared to other forms of energy.

Electrical energy: A form of energy created by small, charged particles known as electrons that usually moves through a wire. A natural example of electrical energy is lightning.

2.3 Types of Energy

Energy comes from many sources. To simplify the expression, the energy sources are classified under two main groups; renewable and nonrenewable. Technological advancement and the sustainability and environmental impact of conventional energy sources, urgent development of producing replenished, and clean energy with sustainable power are required which now arouses interest around the world. Figure 2.7 shows the renewable and nonrenewable energy resources. In general, the nonrenewable energy is divided into four subgroups (based on its energy resources), while the renewable energy sources are categorized under five main subgroups. These energy resources will be discussed more in this section.

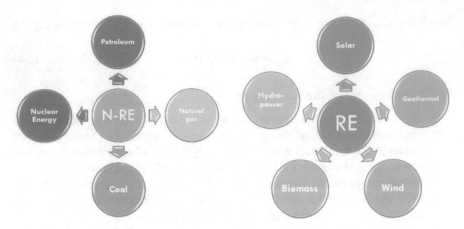

Fig. 2.7 Various types of energy resources

2.3.1 Nonrenewable Energy

Nonrenewable energy (NRE) resources are defined as the sources once they are used up, cannot be easily restored (or replenished). Petroleum, natural gas, coal, and uranium (nuclear power) are the main resources in this field. Most nonrenewable energy resources are fossil fuels—coal, petroleum, and natural gas—i.e., the hydrocarbon-containing natural resources. Carbon is the main element in fossil fuels, and the duration (period) of fossil fuel formation (around 400 million years ago) is called the Carboniferous Period. The formation of all fossil fuels follows similar pathway, before the dinosaurs (millions of years ago), Earth had a different landscape, which covered with seas and forests, appropriate for fungi, algae, and plankton growth. These active biological species absorbed sunlight and created energy among the O_2 to CO_2 cycle (photosynthesis). The dead biological sources drifted to the bottom of the surface water (sea, lake, etc.), containing stored energy. Over time, the hard layer of lithosphere (rocks and other sediments) cover them and create appropriate conditions (high temperature, and pressure), where they convert to fossil fuels.

Crude oil Around 40% of world energy still comes from petroleum. The majority of crude oils (which is also called black oil) are sourced from the Middle East. Crude oil is a mixture of hydrocarbons (either volatile or nonvolatile) that is formed from plants and animals that lived millions of years ago. Crude oil is a fossil fuel and it exists in liquid form in underground pools or reservoirs, within sedimentary rocks, and near the surface in tar (or oil) sands. Sedimentary rocks are rocks containing natural gas and oil, which are formed when minerals and grains from running water merge together. These rocks are relatively porous due to their fusion from such small particles, and this structure allows energy-rich carbon compounds to settle in their pores which can later be extracted in the form of gas or oil. Tar sands (also called oil sands) are a mixture of sand, clay, water, and bitumen. Bitumen is a thick, sticky, black oil that can form naturally in a variety of ways, though it is most commonly formed when a lighter oil is degraded by bacteria. After extraction from the ground, crude oil is usually sent to a refinery where it will be separated into various functional petroleum products. These products are fuels created from the extracted crude oil and other hydrocarbons, kept in natural gas through a process called refinery. Gasoline, jet fuel, waxes, asphalt, petrochemical feedstocks, lubricating oils, and distillates like diesel fuel and heating oil are examples of popular petroleum products.

Petroleum refining is a process involving the conversion of heavy crude oil into lighter petroleum products. The process of refining basically involves the breaking down of crude oil into its many different components, which are then reformed into new products. Petroleum refineries are complex and expensive industrial facilities that carry out four main processes: desalting, separation/distillation, conversion, and treatment (Fig. 2.8).

Natural gas An energy source derived from fossils buried deep in the Earth. Its main component is methane, a compound consisting of one carbon atom and four hydrogen atoms (CH_4). Small amounts of natural gas liquids (NGLs, which are hydrocarbon gas liquids) and non-hydrocarbon gases like carbon dioxide and water vapor are also found in natural gas. We use natural gas as a fuel and to make materials and chemicals. Hundreds of millions of years ago, much like other fossil fuels, the remnants of living things (for example, diatoms) collected in thick layers on ocean floors and in the ground, occasionally mixing with sand, calcium carbonate, and silt. As time progressed, these layers were further buried under rocks, sand, and silt. Due to a combination of pressure and heat, some of these carbon and hydrogen-rich compounds were converted into coal, oil (petroleum), and natural gas. Natural gas from a wellhead is too impure for use; therefore, it is purified in a natural gas processing unit and this pure form is used by consumers (Fig. 2.9).

Coal A combustible black or brownish-black sedimentary rock, containing carbon and hydrocarbons. It is classified as a nonrenewable energy source because it takes millions of years to form. Coal mostly consists of the energy stored by plants in swampy forests that lived hundreds of millions of years ago. Coal is ranked based on its amount of carbon (carbon content) as well as the heat energy it can produce.

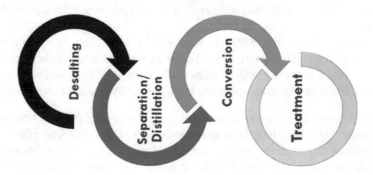

Fig. 2.8 Industrial refinery processes

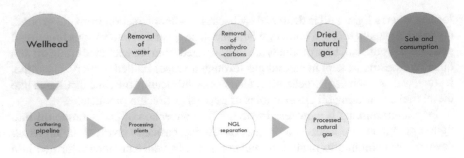

Fig. 2.9 Industrial natural gas processing unit

- Anthracite contains 86–97% carbon and generally has the highest heating value
- Bituminous coal contains 45–86% carbon
- Subbituminous coal typically contains 35–45% carbon
- Lignite contains 25–35% carbon and has the lowest energy content.

Coal can be used in solid form or it can be converted to gaseous, liquid, and solid fuel. The two most common categories of coal conversion are gasification and liquefaction. Gasification of water slurry occurs between 1250 and 1500 °C at a pressure between 350 and 1200 lb/in.^{2}. The chemical addition of hydrogen to coal can liquefy coal and produce synthetic petroleum.

2.3.2 Renewable Energy

Renewable energy (RE) is an energy source that can be easily replenished such as solar, geothermal, wind, biomass, and hydropower. Renewable power is cheap and promising clean energy for today and future life. Harnessing nature power has long been used for various applications such as heating, transportation, lighting, and so on. For example, wind for boats, sun for warming during the day. However, the traditional utilization of RE has been changed over time, when the humans looked for cheaper, more stable, and dirtier energy sources. Nowadays, in many countries, the solar and wind generate without compromising reliability, owing to the many advantages of RE such as lower emissions of carbon and other types of pollution. However, some sources of energy marketed as "renewable" are not really beneficial over the environment due to their negative impact on climate, wildlife, and bioactivity.

Biomass Biomass is vegetative and animal waste (organic matter) that can be converted into useful energy. It contains stored solar energy absorbed from plants during photosynthesis. The chemical energy stored in biomass is released as heat when it is burned. Biomass can be burned directly (solid fuel) or converted to liquid biofuels or biogas that can be burned as fuels (Fig. 2.10). Wood and wood processing wastes,

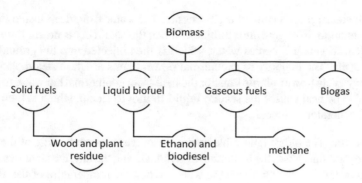

Fig. 2.10 Various types of biomass fuels

agricultural crops and waste materials, domestic food and wood waste, and animal and human sewage are important examples of biomass.

Geothermal Geothermal energy is heat from within the Earth. The word geothermal comes from the Greek words *geo* (Earth) and *therme* (heat). Underground heat in the form of steam, hot water, or hot rocks are used to produce steam has been used traditionally as an energy resource for over a century. The Earth has four major parts or layers:

- A 1500-mile-thick solid iron inner core
- A 1500-mile-thick molten rock (magma) outer core
- A 1800-mile-thick magma and rock mantle surrounding the outer core
- A solid rock crust that is 15–35 miles thick under the continents and 3–5 miles thick under the oceans.

Geothermal power plants utilize both water (hydro) and heat (thermal) resources and need high-temperature (300–700 °F) hydrothermal resources from either hot water or dry steam wells. The hot water or steam powers a turbine that generates electricity. There are three types of geothermal power plants (Fig. 2.11):

- **Dry steam plants** that use steam directly from a geothermal reservoir to drive generator turbines. The first geothermal power plant was built in 1904 in Tuscany, Italy, where natural steam erupted from the Earth.

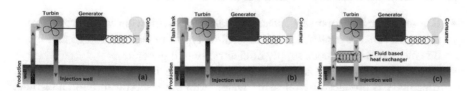

Fig. 2.11 Various types of geothermal power plants **a** dry steam, **b** flash steam, and **c** binary cycle

- **Flash steam plants** convert high-pressure hot water from deep inside the Earth into steam to drive generator turbines. When the steam cools down, it undergoes condensation and becomes water, which is then injected into the ground for the next cycle. The majority of geothermal power plants are flash steam plants.
- **Binary cycle power plants** transfer the heat from geothermal hot water to another liquid. The heat causes the second liquid to turn to steam, which is then used to drive generator turbines.

Wind power The oldest renewable energy resource. Uneven heating of the Earth's surface by the sun causes the formation of wind. During the day, the air above the land heats up faster than the air above the water. When the temperature of the air above land increases, heavier and cooler air moves into take its place—creating wind—and the opposite happens at night. Wind energy can be collected. The blades on wind turbines are used to collect the kinetic energy of wind as it moves over them, creating lift (similar to airplane wings) which turns the blades. The blades are connected to a drive shaft that turns an electric generator, which in turn produces electricity.

One problem with wind turbines is that individually they do not generate a lot of electricity. They are usually needed in large numbers to have an impact on electricity production. A group of wind turbines is called a wind farm. Wind farms require large amounts of space in open areas, but the land can also be used for farming at the same time.

There are two basic types of wind turbines: Horizontal-axis turbines and Vertical-axis turbines (Fig. 2.12). Horizontal-axis turbines are similar to airplane propellers, consisting of three blades. The largest horizontal-axis turbines have blades more than 100 feet long. Longer blades on taller turbines generally produce more energy. The majority of current wind turbines are horizontal-axis turbines. Vertical-axis turbines have a shape similar to an egg-beater, with blades attached to the top and bottom of its vertical rotor.

Hydropower There is a long history of using flowing water to produce mechanical energy. Hydropower was one of the first sources of energy used for electricity generation and is the single largest renewable energy source for electricity generation in many countries. Hydroelectric power is produced from moving water, and the majority of hydroelectricity is produced by large dams. The ranking of each state in annual hydroelectricity generation tends to differ from its ranking in generation capacity because the generation of electricity using hydropower strongly relies on precipitation. There are four major types of hydropower:

- **Run-of-river hydropower**: a facility that channels flowing water from a river through a canal or penstock to spin a turbine, typically with no storage facility. In this method, a continuous supply of electricity will generate.
- **Storage hydropower**: a large system in a dam where the water store in a reservoir. Electricity is produced by releasing water from the reservoir through a turbine, which activates a generator. It has a large amount of storage capacity, in order to operate hydrological inflow for a long period of time.

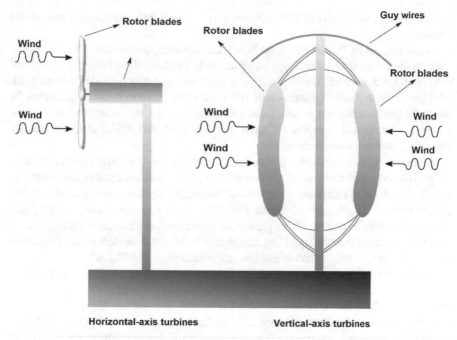

Fig. 2.12 Horizontal-axis and vertical-axis wind turbines

- **Pumped-storage hydropower**: provides peak-load supply, harnessing water which is cycled between a lower and upper reservoir by pumps that use surplus energy from the system at times of low demand.
- **Offshore hydropower**: a less established but growing group of technology that uses tidal currents or the power of waves to generate electricity from seawater. Tidal energy (or tidal power) is a form of hydropower that uses water to create energy. Tidal generators convert the energy produced from wave movement into electric power using tidal energy. There are many methods used to generate this energy, but tidal stream generators or tidal energy converters (TEC) are the most popular due to their low cost and environmental friendliness. Similar to the way wind powers wind turbines, the kinetic energy of moving tides power tidal stream generators.

Solar power It only takes 18 days of sunshine on Earth to produce the amount of energy equivalent to the energy contained in Earth's entire reserves of oil, coal, and natural gas. The sun's energy is around $1300 \, \text{W/m}^2$ outside the atmosphere and once it reaches the atmosphere, two-thirds of it is radiated towards Earth while the rest is reflected back into space. Solar cells (or photocells) turn light energy from the sun directly into current electricity. Manufacturing solar cells are very expensive and require the use of highly toxic materials; however, once the solar cell is built, it produces no pollution and requires little maintenance. Photovoltaic (PV) panels and concentrating solar power (CSP) facilities harness sunlight and convert it into useful electricity.

In the year 1839, French physicist Edmond Becquerel discovered that some materials produced sparks of electricity when sunlight hit them. This property was called the "photoelectric effect", and researchers soon realized that it could be harnessed. This gave rise to the first photovoltaic (PV) cells, which were made of selenium. In the early 1950s, scientists at Bell Labs further studied the technology using silicon, where they managed to create PV cells that could convert 4% of the solar energy present in sunlight into electricity.

An external circuit provides a path for electrons to travel from the n-type layer to the p-type layer, and this flow of electrons through the circuit generates the electricity supply. Most PV systems are comprised of individual square cells measuring a few inches per side. Each cell generates very little power (a few watts), so they are grouped together as panels. These panels are either used as separate units or grouped into larger arrays. An "N" type semiconductor is an electron-rich phase (negative) while a "P" type semiconductor has few or no electrons (Fig. 2.13a).

Unlike photovoltaic (PV) systems, which use sunlight to generate electricity, concentrating solar power (CSP) systems generate electricity using the sun's heat (Fig. 2.13b). The CSP is also called solar thermal power. Various designs of CSP are used, but the most common CSP technology consists of long, curved mirrors that concentrate sunlight on a liquid (generally oil) inside a tube that runs parallel to the mirror. The liquid, which is heated to a temperature of around 300 °C, then

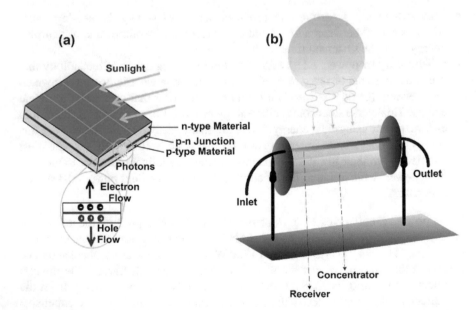

Fig. 2.13 Schematic representation of **a** photovoltaic panel, and **b** concentrating solar power (CSP)

enters a central collector to produce steam which drives an electric turbine. This mechanism is called a Parabolic Trough.

2.4 Summary

Energy is the ability to do work. Energy is a quantitative attribute that must be shifted to an object in order for work to be carried out or for an object to be heated. Energy is a conserved quantity; the Law of Conservation of Energy states that energy can be converted to other forms, but not created or destroyed. Energy is a key element of interaction between nature and society and is crucial for the environment and sustainable development.

Potential energy is stored energy and the energy of position while working (kinetic) energy is the motion of waves, electrons, atoms, molecules, substances, and objects.

Nonrenewable energy sources are responsible for about 90% of world energy consumption. Biomass such as wood, biofuels, and biomass waste, is the biggest renewable energy source and accounts for almost half of the world's renewable energy consumption.

Nonrenewable energy seems abundant, safe, and ample for future generations. However, there are several arguments for and against nonrenewable energy. Renewable energy is a clean source of energy that can easily be replenished, and it is safe for the environment and our health. However, there are also some disadvantages to renewable energy. The advantages and disadvantages of renewable energy are summarized in Table 2.1.

Owing to some disadvantages of renewable energy, a more convenient form of energy, which can transform from primary energy sources through energy conversion processes, is urgently required. Secondary energy is one of the solutions to the primary energy challenges. Secondary energy sources refer to as energy carriers, since they move energy in a useable form from one place to another. One of the most well-known energy carriers is hydrogen. However, all energy sources, after production, must be stored.

Table 2.1 Advantages and disadvantages of renewable energy

Advantage	Disadvantage
Easily regenerated	Weathered dependency
Boost economic growth	High installation cost
Easily available	Pollution caused from some sources either directly or from the waste such as noise, harmful chemicals, etc.
Support environment	Low efficiency
Low maintenance cost	

Acknowledgements Ministry   of   Education   Malaysia   TRGS   research   grant (203/PTEKIND/67610003), and Universiti Sains Malaysia Postdoctoral Scheme.

References

D.J. Davidson, Exnovating for a renewable energy transition. Nat. Energy (2019). https://doi.org/ 10.1038/s41560-019-0369-3

S. Kuppusamy, N.R. Maddela, M. Mallavarapu, K. Venkateswarlu, Total petroleum hydrocarbons. Environmental fate, toxicity, and remediation. Environ. Anal. Technol. Refin. Ind. (2020). https:// doi.org/10.1007/978-3-030-24035-6

S. Kuppusamy, N.R. Maddela, M. Megharaj, K. Venkateswarlu, S. Kuppusamy, N.R. Maddela, M. Megharaj, K. Venkateswarlu, Fate of total petroleum hydrocarbons in the environment, in *Total Petroleum Hydrocarbons* (Springer, Cham, 2020b), pp. 57–77. https://doi.org/10.1007/978-3- 030-24035-6_3

M.L. Latash, V.M. Zatsiorsky, Mechanical work and energy, in *Biomechanics and Motor Control* (Elsevier, San Diego, 2016), pp. 63–82. https://doi.org/10.1016/b978-0-12-800384-8.00004-1

G. Livadiotis, Phase space kappa distributions with potential energy, in *Kappa Distributions: Theory and Applications in Plasmas* (Elsevier Inc., San Diego, 2017), pp. 105–176. https://doi.org/10. 1016/B978-0-12-804638-8.00003-6

B. Mullan, J. Haqq-Misra, Population growth, energy use, and the implications for the search for extraterrestrial intelligence. Futures **106**, 4–17 (2019). https://doi.org/10.1016/j.futures.2018. 06.009

S. Şahin, H. Mehmet Şahin, Nuclear energy, in *Comprehensive Energy Systems* (Elsevier Inc., New York, 2018), pp. 795–849. https://doi.org/10.1016/B978-0-12-809597-3.00122-X

S. Şahin, Y. Wu, Fission energy production, in *Comprehensive Energy Systems* (Elsevier Inc., Amsterdam, 2018), pp. 590–637. https://doi.org/10.1016/B978-0-12-809597-3.00331-X

M. Schmal, *Chemical Reaction Engineering: Essentials, Exercises and Examples* (CRC Press, Boca Raton, 2014)

Y. Wu, S. Şahin, Fusion energy production, in *Comprehensive Energy Systems* (Elsevier Inc., Amsterdam, 2018), pp. 538–589. https://doi.org/10.1016/B978-0-12-809597-3.00330-8

J.W. York, Energy and momentum of the gravitational field, in *Essays in General Relativity* (Elsevier, New York, 1980), pp. 39–58. https://doi.org/10.1016/b978-0-12-691380-4.50010-1

Chapter 3
Energy Storage Systems

Francis Thomas Bacon O.B.E.
FR. Eng. F.R.S., an English
engineer who developed the
first practical
hydrogen–oxygen fuel cell

Abstract Various types of storage technologies have been created so that the grid can achieve daily energy requirements. Ever since electricity was discovered, mankind has constantly searched for effective ways to store energy so that it can be used instantly when required. For the past century, technological advancement and shifting energy requirements have forced the energy storage industry to adapt and evolve. This chapter discusses an overview and types of energy storage systems.

3.1 Overview

Energy storage systems are generally a system or technology that contains electric energy so it can later be distributed through a network when required. For the past few decades, the energy storage industry has continuously adapted and trans-

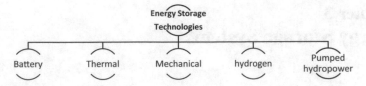

Fig. 3.1 Energy storage technologies

formed to meet energy requirements and as a result, many different types of storage technologies have been created to contain energy temporarily so it can be used when required. Storage devices balance a network's load and generated power using frequency regulation.

Enhanced energy storage can provide multiple benefits:

- Enhanced power quality and electricity delivery to consumers
- Better stability and dependability of distribution and transmission systems
- Greater usage of existing machinery (expensive upgrades are not required)
- Better availability and market value of generation distribution sources
- Greater value for the production of renewable energy
- Cheaper costs due to the deferral of transmission and capacity fees.

Based on diverse approaches currently have deployed, energy storage technologies can be classified into five main categories (Fig. 3.1).

As mentioned before, an energy storage system can store energy to produce electricity and discharge it, depending on the demand or cost benefits. Criteria effectiveness for a typical energy storage system include (Das et al. 2018):

Dispatchability—A dispatchable source of electricity refers to electrical demand fluctuations, which can occur repeatedly and relatively quickly, for instance, power plants. Dispatchability is the ability to adjust output power to the electrical grid on demand. Most nonrenewable energy sources (oil, coal, natural gas) are dispatchable, while many renewable energy sources are non-dispatchable, such as wind power or solar power. The former can change the electricity demands of the population, while the latter can only generate electricity while their energy flow is input. The electricity demand fluctuates on different cycles because of the changes in domestic and industrial loads, and environmental factors.

Interruptibility—Reactivity to the intermittency of renewable energy is known as interruptibility. Fossil fuel supplies are often unstable, while solar and wind energy are season-dependent behaviors of biomass and hydropower. Interruptibility provides a rapid and efficient response to the electric system needs according to technical and economic criteria. Some companies in the world provide a service on interruptibility, which is a response to power management to large consumers mainly large-scale industry.

Efficiency—An energy storage system is a system with the ability of storing energy, during high electricity production periods and return it to consumption in a period of time. This system is characterized by energy storage capacity ($E_{cap.}$).

The hours of the installation's energy autonomy, the maximum discharge depth, as well as the efficiency of energy transformation of the energy storage system. Generally, efficiency can be defined as the capacity to reclaim and reuse energy that would otherwise go to waste.

3.2 Introduction

Hydrogen, which is known as the tenth most abundant element on Earth, has very rich chemistry despite its simple atomic structure (He et al. 2016). Its reaction profiles are particularly interesting in terms of fundamental chemistry and applications, especially in terms of "energy". Hydrogen molecule has a few electrons; therefore, the intermolecular forces between H–H molecules are weak. It is known that at around 1 atm pressure, the gas condenses to a liquid. Applying an electric discharge through H_2 gas at low pressure result in molecular dissociation, ionization, recombination, and formation of plasma-add-on (Of et al. 2019). H-atom has high ionization energy (1312 kJ mol^{-1}) and a partially positive electron affinity (73 kJ mol^{-1}) (Atkins et al. 2006). These abilities can cause to formation of binary hydrogen compound such as molecular hydrides, saline hydrides, and metallic hydrides (Mohtadi and Orimo 2017). Hydrogen is an important raw material for chemical industries and fuel in mobile applications (Schlapbach and Züttel 2001). In the small scales, hydrogen can be produced by the reactions of electropositive elements in an alkaline or acidic media, by electrolysis, or by thermolysis of water. Direct thermolysis of water requires very high temperatures of around 4000 °C, which is much higher than the threshold value (Suib et al. 2013).

Hydrogen molecule (H_2) can be dissociated either by homolytic or heterolytic processes on an activated surface containing metals (M)/metal oxides (MO)/mixed metal oxides (MMO) (Chen et al. 2018). Homolytic dissociation is induced by adsorption on the activated metallic substrate as Eq. 3.1

$$H_2 + 2M \leftrightarrow 2M - H, \qquad (3.1)$$

while heterolytic dissociation into H^+ and H^- occurs on a heteroatom substrate like metal oxides or mixed metal oxides Eq. 3.2

$$H_2 + M - O \leftrightarrow^- H - M - O - H^+. \qquad (3.2)$$

Onboard hydrogen storage is a major obstacle to the future energy carrier. Interest in materials-based hydrogen storage has been started by the researcher of the Brookhaven National Laboratory in the USA in the 1960s, and accidentally discovered in 1969 in a research on the rare-earth AB_5 intermetallic materials (Cordtz et al. n.d.). Fossil fuels supply 70% of today's electrical energy, with nuclear and hydropower supplying the other 30% and other renewable energy technologies supplying around 3% (Yang et al. 2011). As of late, the government has been forced into looking at energy production from renewable sources due to resource constraints and environmental concerns (Pires et al. 2014). Wind and solar power are readily available and highly abundant, but these energy sources lack reliability and consistency (Caponigro 2011). Hence, alternatively, hydrogen is foreseen to be the future in renewable energy sources.

However, the main challenge in the storage of hydrogen is to find a new class of materials that can reversibly store hydrogen at high rates under reasonable temperatures, pressure and cost conditions. In the traditional compressed storage, high-pressure gaseous refrigerate to form liquid H_2 which requires a large area with considerable energy and containment costs.

3.3 Battery

A battery uses a reversible chemical reaction to store electrical energy. Energy is usually produced by a renewable energy (RE) source such as PV, wind, or hydro, and the battery then stores it in case of a time of reduced or no RE production. Most batteries used in RE systems run on the same electrochemical reactions that occur in the lead-acid battery of your car, but its main difference is that they are specially designed for deep cycling and can store around ten to a hundred times more energy. This, however, does not ensure consistent performance, and a backup power source will always come in handy if your batteries become discharged due to reduced RE in the system or energy overconsumption.

There are various types of batteries produced for RE systems, and it is crucial to select the right battery according to the situation and your finances. The two most common sizes of batteries are the L-16 and golf cart. Most RE system batteries can last five to ten years with proper care, but there are more expensive batteries of a higher quality that can have a lifetime of even ten to twenty years.

Battery capacity is calculated in amp-hours. One amp-hour is generally defined as the steady drawing of 1 A for 1 h or 2 A for 30 min. A typical 12-volt system can have a battery capacity of around 800 amp-hours, which means that it can draw 100 A for 8 h when completely discharged and beginning from a state of full charge. This is equal to 1200 W for 8 h (watts = volts × amps), which uses about the same power as a small hairdryer running for 8 h.

On its most basic level, a battery is a device consisting of one or more electrochemical cells that convert stored chemical energy into electrical energy. Each cell contains a positive terminal, or cathode, and a negative terminal, or anode. Electrolytes allow ions to move between the electrodes and terminals, which allows current to flow out of the battery to perform work. The solid-state batteries (a battery with solid electrodes and a solid electrolyte) are found many advantages over the liquid or polymer gel electrolytes batteries. For example, liquid-lithium-ion batteries can leak or ignite rapidly if they become overheated, while the solid-state lithium-ion batteries are safe

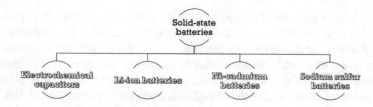

Fig. 3.2 various types of solid-state batteries

to use. However, these classes of batteries are still in their beginning stages and more research is required for their developments. Figure 3.2 represents the most important commercialized solid-state batteries.

For instance, a Li-ion battery is a type of rechargeable battery. In LIB, lithium ions move from the negative electrode to the positive electrode during discharge and back when charging. In rechargeable LIB, intercalated lithium compound is used. Here, a substrate like carbon materials (graphite, CNT, etc.) carries the metal oxides. In non-rechargeable LIB, metallic lithium used in a non-rechargeable lithium battery (Placke et al. 2017). Figure 3.3 illustrates the details of a modern rechargeable Li-ion battery.

Fig. 3.3 A modern rechargeable Li-ion battery

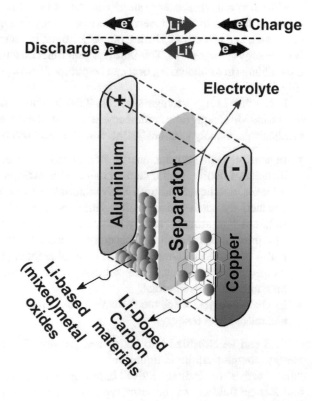

A flow battery is a type of rechargeable battery where rechargeability is provided by two chemical components dissolved in liquids contained within the system and most commonly separated by a membrane. One of the biggest advantages of flow batteries is that they can be almost instantly recharged by replacing the electrolyte. Redox flow (RB), Ni-Cd flow (ICB), vanadium redox flow (VRB), and zinc-bromide flow (ZBB) batteries are some examples of flow batteries. For example, in VRB, vanadium ions in different oxidation states can store chemical potential energy. Vanadium exists in four different oxidation states. The electrodes in a VRB cell are generally carbon materials. The electrolyte in the positive half-cells contains V^{5+} and V^{4+}. The electrolyte in the negative half-cells are V^{3+} and V^{2+} ions.

3.4   Thermal Energy Storage System

Thermal energy storage (TES) is an environmentally friendly storage technology, and economically feasible, where the thermal energy stocks either by heating or cooling in a storage medium. This amount of stored energy can be further used for continuous heating and cooling. In addition, it can be used as a power generation in buildings and in industrial processes.

How does a thermal energy storage system work? The system contains six major elements such as thermal receiver, thermal energy storage tanks, steam generator, condenser, and powerline. During off-peak hours, ice is made and stored inside Ice-Bank energy storage tanks. This tank has dual functions; the stored ice can cool down the building (in a condenser), or it can be pumped into a generator to generate power (Fig. 3.4).

Like other energy storage systems, TES is also described in terms of capacity, charge/discharge time, efficiency, and cost. There are a few types of storage mechanisms, namely sensible, latent, and chemical reactions:

- In **sensible-type storage**, energy storage is carried out through the temperature increase of liquid or solid storage media (for example, concrete, oils, sand-rock minerals, and liquid sodium). Such materials are cheaper and have high thermal conductivity, but they also require a large system size due to their low heat capacity.
- In **latent-type storage**, energy is stored (and released) during a change in phase. This provides a larger storage capacity compared to the sensible-type storage, but it also lacks thermal conductivity. It is suitable for applications that have working temperature constraints, as the solid-liquid phase change process of eutectic or pure materials is isothermal.
- In **chemical-type TES**, the formation, and breaking of chemical bonds absorb and release heat (energy).

TES can be classified into active and passive TES depending on whether the energy storage medium is liquid or solid. Active TES can be further classified as "direct active" or "indirect active" based on whether the heat transfer fluid (HTF) and storage fluid are of the same type, or if another type of HTF is necessary to

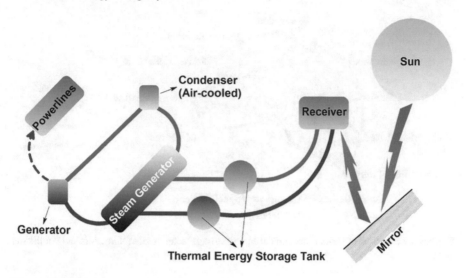

Fig. 3.4 Thermal energy storage system

extract heat from a solar field. Passive thermal energy systems can easily improve available natural heat energy sources to keep conditions in buildings comfortable while also reducing the usage of mechanically assisted cooling or heating systems.

3.5   Mechanical Energy Storage

Physical connection between a flywheel and the driven wheels is known as mechanical energy storage, which generally occurs through a continuously variable transmission (CVT). In modern technology, the conversion of electrical energy into mechanical energy, and their storage profiles in terms of kinetic or potential energies is called mechanical energy storage, for example, flywheels and pumped-hydro. An electric motor is responsible for electrical to mechanical energy conversion, since discharging the stored energy.

Mechanical energy storage systems can be classified in terms of the technologies used. Pumped Heat Electrical Storage (PHES), compressed air energy storage (CAES), flywheel energy storage (FES), pumped-hydro energy storage (PHES), compressed air energy storage (CAES), etc. are the most mature technologies in mechanical energy storage possessing. These technologies provide the greatest energy capacities and power ratings, with the longest lifetime and a cheap cost. However, FES, PHES, and CAES systems need massive storage reservoirs to generate a specific amount of energy and power. Figure 3.5 compares the potential characteristics of mechanical energy storage technologies.

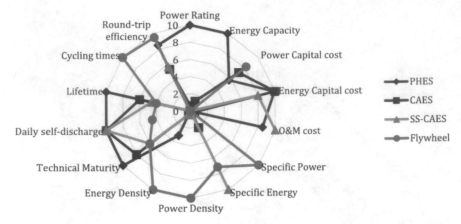

Fig. 3.5 Comparison of various mechanical energy storage technologies (Nikolaidis and Poullikkas 2018)

3.6 Hydrogen Energy Storage

Hydrogen energy storage is a form of energy storage in which the hydrogen can be stored. This energy can be further released as fuel in a combustion engine or a fuel cell. Electrolysis of water is one of the easiest ways to generate hydrogen, with relatively high efficiency (Stamenkovic et al. 2016). The hydrogen stores in underground caverns either for large-scale energy storage or in containers for smaller scale storage. Hydrogen can be used as fuel for piston engines (internal combustion engine), gas turbines, or hydrogen fuel cells. The electrochemical hydrogen storage in fuel cells offers the best efficiency.

- **Hydrogen-based internal combustion engine**—Hydrogen can be used as a fuel in conventional spark-ignition engines. Upon combustion of hydrogen, water vapor comes out which further goes back to the biosphere. The hydrogen-based engine efficiency is high as compared to petroleum-based fuels. In the presence of hydrogen, flame expands rapidly from the kernel of ignition. Apart from the performance of a typical fuel induction in the hydrogen engine, there is a proper controlling over the tendency of its backfire.
- **Hydrogen gas turbine**—Hydrogen gas turbines can be used in critical industries such as power generation, oil and gas, process plants, aviation, etc. In order to decarbonize this specific field, the hydrogen gas can be used, due to its reactivity. The high reactivity of hydrogen gas leads to a high laminar burning velocity. In fact, the hydrogen can extend the flammability limits and flame propagation of turbine. As a result, efficient combustion, with low emission of pollutants can occur.
- **Hydrogen fuel cell**—fuel cell operates as a battery, i.e., it generates electricity from an electrochemical reaction. A fuel cell works with three main elements;

hydrogen, oxygen, and electrolyte. External chemical energy reacts with hydrogen and oxygen. The product of this reaction is water. In a fuel cell, there is no combustion involved. The oxidation of hydrogen happens electrochemically; in this process, hydrogen atoms will react with oxygen atoms and form water. Electrons released through this process flow as an electric current through an external circuit. Solid Oxide Fuel Cells (SOFC), Proton Exchange Membrane Fuel Cells (PEMFC), Alkaline Fuel Cells, Direct Methanol Fuel Cells (DMFC), Phosphoric Acid Fuel Cells (PAFC), and Molten Carbonate Fuel Cells (MCFC) are the most abundant used fuel cells. Table 3.1 compares some features of these fuel cells.

Why hydrogen energy storage is important? Hydrogen has the potential to replace fossil fuels in combustion profiles, making it crucial for the hydrogen economy. Unfortunately, the overall efficiency of hydrogen storage system is lower than

Table 3.1 Important features of various types of fuel cells

	Electrolyte	Operation temperature (°C)	Efficiency (%)	Application	Deficiency
SOFC	Solid ceramic	800–1000	50–60	• Auxiliary power units in vehicles • Stationary power generation • Heat engine energy recovery devices	• Vulnerability to sulfur
PEMFC	Water-based acidic polymer	>100	50	• Road transport application	• Low operation temperature • Expensive platinum catalyst material
AFC	Alkaline solution	23–70	60	• Shuttles • Spacecraft	• Sensitive to carbon dioxide
DMFC	Polymer	60–130	40–50	• Mobile electronic devices or chargers • Portable power packs • Electric city cars	• Expensive platinum-ruthenium catalyst
PAFC	Liquid phosphoric acid	150–200	70	• Stationary applications	• CO sensitive • Expensive platinum catalyst material
MCFC	Molten carbonate salt mixture suspended in a porous, chemically inert ceramic lithium aluminum oxide ($LiAlO_2$) matrix	650	60–85	• Electrical utility • Industrial, and military applications	• Low power density • Aggressiveness of the electrolyte

other storage technologies. However, despite this deficiency, the interest in this system is growing due to its high storage capacity compared to batteries or pumped-hydro. On the other hand, the diversity of potential supply sources is important in order to utilize hydrogen as a promising energy carrier.

Alkaline electrolysis is a technology for a large hydrogen storage system. In addition, Proton Exchange Membrane (PEM) electrolyzers are more flexible, more efficient, and more recyclable for small-decentralized solutions. The efficiency of both technologies is about 70%. High-temperature PEM is a new technique offers an enhanced efficiency of around 90%.

Though it is Earth's most common element, hydrogen can be early found in its pure form. This would imply that hydrogen would require extraction from its compound for usage. Small amounts of hydrogen can be stored either in pressurized vessels or in solid-state materials (porous materials) with a very high density. Various resources from renewable energy sources and fossil fuels can be used to split water for the production of hydrogen.

The government of several European and American countries offers integrated hydrogen solutions and hydrogen regulation for the supply of electric power to small sites, such as the U.S. Department of Energy (DOE), and hydrogen Europe. These associations identify the legislation and regulations relevant to fuel cell and hydrogen applications. In addition, they work on the legal barriers to hydrogen commercialization.

3.7 Fuel Cell

Fuel cell uses the chemical energy of hydrogen or another fuel to produce electricity. In the field of fuel cell, routine reliable synthesis of self-assembled materials with tunable functionalities is urgently required. Novel functional nanostructures and polymer-based nanocomposites/hybrids are recently developed for fuel cell application.

There are various types of fuel cells, all consist of an anode, a cathode, and an electrolyte that allows positively-charged hydrogen ions (or protons) to move between the two sides of the fuel cell (Fig. 3.6). Among them, proton exchange membrane fuel cell (PEMFC) and solid-state fuel cell (SFC) have been achieved a great interest due to high working temperature, high qualified power, great efficiency (cell and system), and good chemical stability. In a typical fuel cell, the anode and cathode contain catalysts that cause the fuel to undergo oxidation reactions that generate positively-charged hydrogen ions and electrons. The hydrogen ions are drawn through the electrolyte after the reaction. At the same time, electrons are drawn from the anode to the cathode through an external circuit, producing direct current electricity, and in cathode, hydrogen ions, electrons, and oxygen react to form water (Cano et al. 2018).

If hydrogen is the fuel, electricity, water, and heat are the only products. Fuel cells can provide power for systems as large as a utility power station and as small as a

Fig. 3.6 A fuel cell unit

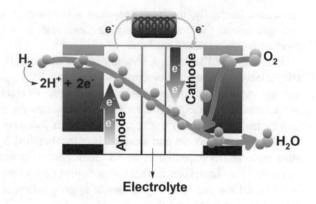

laptop computer. Various applications including transportation, material handling, stationary, portable, and emergency backup power applications. The conversion efficiency of a typical fuel cell is up to 60%, much higher than conventional combustion technologies. No harmful emissions, the only emission is water. No noise pollution.

Polymer electrolyte membrane (PEM) fuel cells are the current focus of research for fuel cell vehicle applications (Wang et al. 2019). PEM fuel cells are made from several layers of different materials. Membrane electrode assembly (MEA), includes the membrane, the catalyst layers, and gas diffusion layers (GDLs). Gaskets, which provide a seal around the MEA to prevent leakage of gases, Bipolar plates, which are used to assemble individual PEM fuel cells into a fuel cell stack and provide channels for the gaseous fuel and air.

High-temperature proton exchange membrane fuel cells (HTPEMFC) offer several advantages, such as high proton conductivity, low permeability to fuel, low electro-osmotic drag coefficient, good chemical/thermal stability, good mechanical properties and low cost.

3.8 Summary

Energy storage technologies are bridging the gap between three factors: power demand, quality of power supplied, and reliability. These technologies are highly attractive due to their applications in large-scale generation, transmission, and network distribution systems. Though a variety of energy storage technologies are developed for everyday lives, however, each technology performs in with its inherent storage characteristics.

Hydrogen energy storage systems, owing to their high-energy density, high capacity, increased storage benefits, durability, reliability, energy conservation, and environmental safety enable them to be preferred in growing energy requirements.

Prior to hydrogen storage, hydrogen should first supply. Hydrogen can be derived from either hydrogen-rich nonrenewable energy resources, or renewable energy resources. Currently, the most prevalent and low-cost hydrogen derives from natural gas among the process of reformation. But, how feasible is future hydrogen production from fossil fuels?

Electrolysis of water is a solution to the above pollution! It is an environmentally friendly method. By electrolysis, it is possible to generate hydrogen from any energy sources but requires substantial amounts of electricity. It is an ideal hydrogen economy if anticipated the electricity required for electrolysis from renewable energy resources. In the next chapter, we will focus on the most recent technique in hydrogen production and storage. Electrochemical hydrogen sorption on solid-state materials is expected to be an ideal system in terms of clean, efficient, and cost-effective. Therefore, further research and development are urgently required, in order to fill the gap in the synthesis of appropriate materials with the potential of hydrogen sorption/desorption.

Acknowledgements Fundamental Research Grant Scheme, Ministry of Education Malaysia (203/PTEKIND/6711574), and Universiti Sains Malaysia Postdoctoral Scheme.

References

P. Atkins, T. Overton, J. Rourke, M. Weller, F. Armstrong, *Inorganic Chemistry*, 4th ed. (W. H. Freeman, USA, 2006)

Z.P. Cano, D. Banham, S. Ye, A. Hintennach, J. Lu, M. Fowler, Z. Chen, Batteries and fuel cells for emerging electric vehicle markets. Nat. Energy (2018). https://doi.org/10.1038/s41560-018-0108-1

M. Caponigro, *Handbook on Renewable Energy Sources* (2011)

B. Chen, Z. Liu, C. Li, Y. Zhu, L. Fu, Y. Wu, T. van Ree, Metal oxides for hydrogen storage, *Metal Oxides in Energy Technologies* (Elsevier Inc., New York, 2018). https://doi.org/10.1016/B978-0-12-811167-3.00009-2

K. Cordtz, T. Gagnon Peter, D. Greenberg, K. Mcnulty, W.K. Snyder, R. Stoutenburgh, Research for our energy future (USA, n.d.)

C.K. Das, O. Bass, G. Kothapalli, T.S. Mahmoud, D. Habibi, *Overview of energy storage systems in distribution networks: placement, sizing, operation, and power quality* (Sustain. Energy Rev., Renew, 2018). https://doi.org/10.1016/j.rser.2018.03.068

T. He, P. Pachfule, H. Wu, Q. Xu, P. Chen, Hydrogen carriers. Nat. Rev. Mater. **1**, 16059 (2016). https://doi.org/10.1038/natrevmats.2016.59

R. Mohtadi, S. Orimo, The renaissance of hydrides as energy materials. Nat. Rev. Mater. **2**, 16091 (2017). https://doi.org/10.1038/natrevmats.2016.91

P. Nikolaidis, A. Poullikkas, Cost metrics of electrical energy storage technologies in potential power system operations. Sustain. Energy Technol. Assess. **25**, 43–59 (2018). https://doi.org/10.1016/j.seta.2017.12.001

D. Of, T. For, H. Pressure, G.A.S.H. Containers, Hydrogen storage and transport technologies, in *Science and Engineering of Hydrogen-Based Energy Technologies* (Elsevier, London, 2019), pp. 221–228. https://doi.org/10.1016/B978-0-12-814251-6.00010-1

V.F. Pires, E. Romero-Cadaval, D. Vinnikov, I. Roasto, J.F. Martins, Power converter interfaces for electrochemical energy storage systems-A review. Energy Convers. Manag. **86**, 453–475 (2014)

T. Placke, R. Kloepsch, S. Dühnen, M. Winter, Lithium ion, lithium metal, and alternative recharge-able battery technologies: the odyssey for high energy density. J. Solid State Electrochem. **21**, 1939–1964 (2017). https://doi.org/10.1007/s10008-017-3610-7

L. Schlapbach, A. Züttel, Hydrogen-storage materials for mobile applications. Nature **414**, 353–358 (2001). https://doi.org/10.1038/35104634

V.R. Stamenkovic, D. Strmcnik, P.P. Lopes, N.M. Markovic, *Energy and fuels from electrochemical interfaces* (Mater, Nat, 2016). https://doi.org/10.1038/nmat4738

S.L. Suib, Y. Kojima, H. Miyaoka, T. Ichikawa, Hydrogen storage materials, in *New and Future Developments in Catalysis* (Elsevier, Amsterdam, 2013), pp. 99–136. https://doi.org/10.1016/B978-0-444-53880-2.00006-5

X.X. Wang, M.T. Swihart, G. Wu, Achievements, challenges and perspectives on cathode catalysts in proton exchange membrane fuel cells for transportation. Nat. Catal. **2**, 578–589 (2019). https://doi.org/10.1038/s41929-019-0304-9

Z. Yang, J. Zhang, M.C.W. Kintner-meyer, X. Lu, D. Choi, J.P. Lemmon, J. Liu, Electrochemi-cal energy storage for green grid. Chem. Rev. **111**, 3577–3613 (2011). https://doi.org/10.1021/cr100290v

Chapter 4
Solid-State Hydrogen Storage Materials

Henry Cavandish (1731–1810), an English natural philosopher in theoretical chemistry. He first discovered the hydrogen as "inflammable air", by reacting zinc metal with hydrochloric acid (1766). In a demonstration to the Royal Society of London, Cavendish applied a spark to hydrogen gas yielding water

Sir William Robert Grove (1811–1896), a Welsh scientist. He is one of the first scientists who expressed the general theory of the conservation of energy, and fuel cell technology (1845), by creating a "gas battery." He earned the title "Father of the Fuel Cell" for his achievement

Abstract Hydrogen is an ideal candidate to fuel as "future energy needs". Hydrogen is a light (Mw = 2.016 g mol^{-1}), abundant, and nonpolluting gas. Hydrogen as a fuel can be a promising alternative to fossil fuels; i.e., it enables energy security and takes cares of climate change issue. Hydrogen has a low density of around 0.0899 kg m^{-3} at normal temperature, and pressure (~7% of the density of air), which is the main challenge in its real applications. It means, for example, 1 kg of hydrogen requires an extremely high volume of around 11 m^3. In order to solve this limitation of hydrogen, solid-state hydrogen storage materials are used to store hydrogen efficiently and effectively. In this chapter, an attempt has been developed to provide a comprehensive overview of the recent advances in hydrogen storage materials in terms of capacity, content, efficiency, and mechanism of storage.

4.1 Introduction

Some criteria are expected for selection of solid-state hydrogen storage systems to be adopted as follows:

- Favorable thermodynamics.
- Fast adsorption-desorption kinetics.
- Large extent of storage (high volumetric and gravimetric density).
- Withstand enough cycle number for both adsorption and desorption.
- Enough mechanical strength and durability.
- Appropriate heat transfer medium.

According to the recent classification of the US Department of Energy (DOE), hydrogen has the energy content of about 120 MJ/kg, i.e., three times higher than that of gasoline (~44 MJ/kg) ("DOE Technical Targets for Onboard Hydrogen Storage for Light-Duty Vehicles | Department of Energy," n.d.). It can be stored either physically in the form of gas/liquid or in/on solid-state materials (Fig. 4.1). The gas-based storage requires high-pressure tanks (350–700 bar), and liquid storage requires cryogenic temperatures ($T_{bp} = -252.8$ °C at 1 atm). Hydrogen can also be accumulated by adsorption (on the solid materials) or by absorption (in the solid materials) ("DOE Technical Targets for Hydrogen Storage Systems for Material Handling Equipment | Department of Energy," n.d.).

These materials are required for onboard vehicle, material-handling equipment, and portable power applications. It aims to commercialize hydrogen-fueled vehicle

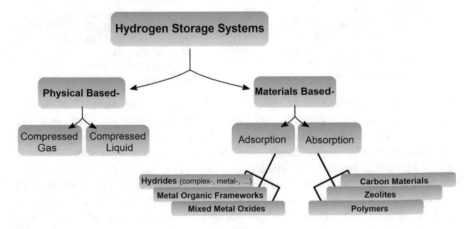

Fig. 4.1 Hydrogen storage systems

Table 4.1 2020s technical system targets for onboard vehicles, materials handling, and portable power equipment published by US-DOE

		Hydrogen capacity 2020 (wt%)	System gravimetric capacity (kWh/kg)	System volumetric capacity (kWh/L)	Storage system cost ($/kWh net)	Min/max delivery temperature (°C)
Vehicles		–	1.5	1.0	10	−40/85
Materials handling		2	–	1.7	15	−40/85
Portable power equipment	Low power (≤2.5 W)	≤1	1.3–1.0	1.7–1.3	0.03–0.4	10/85
	Medium power (>2.5 W–150 W)	>1–50	1.3–1.0	1.7–1.3	0.1–0.5	10/85

platforms according to the customer expectations for range, passenger and cargo space, refueling time, and overall vehicle performance by coming 2020s. This target should meet a 1.5 kWh/kg system (4.5 wt% hydrogen) or 1.0 kWh/L system (0.030 kg hydrogen/L) with $10/kWh ($333/kg stored hydrogen capacity) (The Fuel Cell Technologies, n.d.).

Though many solid-state materials have been manufactured for hydrogen sorption, what are the challenges? The DOE reported that high-density hydrogen storage remains a significant challenge for stationary, portable and transportation applications. The recent invention in hydrogen storage systems requires large-volume systems. It is required a cruising range of about 300 miles with a fast and facial refueling. Howbeit, light-duty hydrogen fuel cell electric vehicles (FCEVs) are available in the market, however, they have less impact on larger vehicles (Section 2015).

The technical targets for a standard hydrogen storage system, including automotive, material handling and portable power, are reported by the US Department of Energy (US-DOE). The results are summarized in Table 4.1. These targets are critical for development, and demonstration planning of future hydrogen storage systems.

4.2 Hydrogen Storage Materials

Solid-state hydrogen storage is one solution to all the above challenges. Materials under investigation include organic polymers, metal–organic frameworks (MOFs), composites/hybrids, alloys, and hydrides (metal-, boro-, and complex-), metal oxides and mixed metal oxides, clay and zeolites, and carbon materials (CNT, graphene). Hydrogen storage can be stored *via* various storage methods currently being investigated,

- high-pressure gas cylinders and liquid hydrogen,
- the physisorption of hydrogen on materials with a high specific surface area,

- hydrogen intercalation in solid-state materials such as metals and complex hydrides,
- storage of hydrogen-based on metals and water,

and what are the goals? The main goals are:

- to pack hydrogen in a small area (close packing)
- to enhance volumetric density
- to increase the discharge capacity
- to harvest charge–discharge efficiency

and what are the challenges? At ambient temperature and atmospheric pressure, 1 kg of the gas has a volume of 11 m^3. This huge volumetric amount of gas must pack in a small area to increase hydrogen density. Compressing the gas, decreasing the temperature below the critical temperature, and/or repulsion reduced by the interaction of hydrogen with other materials are the only available solutions.

4.3   Hydrides

The term "hydride" is used for binary compounds that hydrogen forms with other elements of the periodic table (E_x-H_y). Hydrides are generally classified into three major groups as covalent, ionic, and metallic hydrides.

- Covalent hydrides (Complex hydrides): the hydrogen atom reacts with one or more nonmetals to form hydrides. The hydrogen covalently bonds to a more electropositive element by sharing electron pairs. These hydrides can be volatile (gas) or nonvolatile (liquid), BeH_2, BH_3, AlH_3 are some examples of covalent hydrides. Though BH_3 and AlH_3 are very unstable, they form complex anions in combination with the hydride anion. The common examples are tetrahydridoborate ($NaBH_4$) and lithium tetrahydridoaluminate ($LiAlH_4$).
- Ionic hydrides (IHs): hydrogen compounds of the most electropositive metals (mostly alkali and alkaline earth metals) can form ionic hydrides via liberation of H_2 in contact with Brønsted acids and transfer H- to electrophiles. This class of hydrides is also known as "saline" hydrides or "pseudohalides". The ionic hydrides of the heavier elements react violently with water to generate hydrogen as Eq. 4.1:

$$MgH_2(s) + 2H_2O(l) \rightarrow Mg(OH)_2(s) + 2H_2(g). \tag{4.1}$$

- Metal hydrides (MHs): metal hydrides, which is also known as "interstitial" hydrides, can produce *via* interaction of hydrogen with transition metals. Their unique structures enable MHs to be nonstoichiometric, with variable composition. Based on this idea, a crystal lattice can be proposed in which the H-atoms can fill in between the lattice and not in a definite-ordered filling. The reaction of MHs with water can produce H_2 molecules (Eq. 4.2):

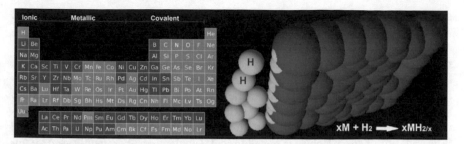

Fig. 4.2 Possibility of elemental hydrogen bonds and the general mechanism of hydrogen storage in hydride forming materials

$$MH_x(s) + nH_2O(l) \rightarrow M(OH)_n(s) + xH_2(g). \tag{4.2}$$

Various hydride based solid materials are currently being considered as hydrogen storages media (Mohtadi and Orimo 2017). These materials are assigned by US-DOE as a target for future hydrogen storage. These materials, based on the previous classification, can be categorized into conventional metal hydrides (interstitial metal hydrides, MgH_2), complex hydrides (Mg_2FeH_6, metal alanates, metal borohydrides, metal-N-H composites), and chemical hydrides (ammonia borane, alane) (Paul-Boncour 2018). In this section, some important metal hydrides (according to US-DOE) will be discussed. A large number of elements can form hydrides via ionic bonds with hydrogen, metallic bonds, and covalent bonds. Figure 4.2 illustrates the elements in hydrogen bonds and the hydrogen storage mechanism of the typical M-hydrides (M: Metal-, complex-, Boro-). In the gas phase, the reaction mechanisms are in five steps: (a) non-dissociative surface physisorption (desorption); (b) surface dissociative chemisorption (recombination); (c) surface absorption (desorption); (d) transport of hydrogen ad-atoms from subsurface to bulk regions by diffusion; and (e) phase transformation leads to the precipitation (dissolution).

4.3.1 Metal Hydrides

Metallic hydrides (MHs) are nonstoichiometric, and conductive solids which can form with many d- and f-block elements. Saline, intermediate, metallic, molecular binary metal hydrides are reported, however, some d-block elements such as iron and ruthenium do not form binary hydrides (Hirscher 2010). These elements are known to be effective for formation of metal complexes. Metal hydrides are known as source of chemical hydrogen sorption. Since the 1970s, a number of alloys have been designed and formulated as an efficient hydrogen storage material, including AB_5 (Young et al. 2015), AB_3 (Latroche 2018), A_2B_7 (Zhang et al. 2014a, b), AB_2 (Zhang et al. 2016), AB (Han et al. 2001), and A_2B (Maeland and Libowitz 1980). In these compounds, A is an absorbing hydrogen metallic element and B represents a weak hydrogen absorber based metallic element but high catalytic activities. B element plays an important role in the hydrogen dissociation process ($H_2 \rightarrow 2H$).

M–H bonds can be either nucleophilic or electrophilic. The nature of the central metal and the reaction conditions govern the properties of MHs. As mentioned before, hydricity refers to the tendency of a hydride ligand to depart as H^-. Hydricities predict from the electronic and steric properties of the metal center. MHs can be synthesized via metal protonation, oxidative addition of H_2, addition of nucleophilic main-group hydrides (borohydrides, aluminum hydrides, and silanes), and β-hydride elimination. However, metallurgical techniques are commonly used to synthesize metal hydrides, and basic chemical operations are used to produce complex hydrides. The process includes heat treatment (for homogenization) and activation (via cyclic hydriding/dehydriding). Other methods such as thin film deposition, sol-gel processing and high-energy mechanical ball milling (for alloying) are used for production of metallic alloys.

Practically, a large variety of alloys and MHs are used in hydrogen storage systems. Though MHs have large volumetric densities of around 115 kg/m^3 (for the reference material LaNi$_5$ compound), however, suffer from low gravimetric densities, as well as high enthalpy of formation (Millet 2014). Hydrogen production metal hydrides occur by reaction with many metals and metal alloys. However, this process can be reversed, by liberating hydrogen and regenerating the metal or alloy. The metal hydrides have an important advantage over hydrogen physisorption (which is considered as hydrogen storage). The compounds contain hydrogen and light elements (such as Li, Be, Na, Mg, B, and Al) offer promising materials in this field, particularly with H:M ratios of two or more. Hydrogen can be incorporated into metals or intermetallic from either an electrolytic solution or from the gaseous phase (Fig. 4.3).

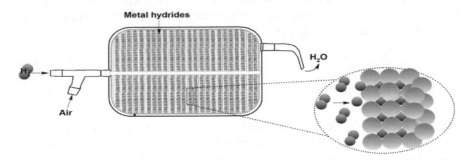

Fig. 4.3 Electrochemical hydrogen storage setup

4.3.2 Complex Hydrides

Complex hydrides are salt-like substances. They form against ionic bonding between a positive metal ion with a molecular anion containing the hydride. Complex hydrides contain stable solids, and convenient to handle (Ley et al. 2014). This class of hydrogen storage materials consists of an electropositive counter ion and a coordination complex, where the hydrogen can covalently bond (Bonyhady et al. 2018). Complex hydrides, tetrahydridoaluminates (AlH_4^-), amides (NH_2^-), and boronates (BH_4^-), contain high amount of hydrogen [26]. These are known as high gravimetric capacitors with high hydrogen gravimetric densities. Complex hydrides are highlighted (by DOE) as an ultimate target for hydrogen sorption and classified into two groups; group I involves simple salt-based hydrides, and group II possesses transition metal complex hydrides (Atkins et al. 2006). The interstitial metal hydrides (MH) formed by the heavier d- and f-block metals and some alloys have received significant attention due to reversible hydrogen storage at moderate conditions (Mohtadi and Orimo 2017).

The as mentioned complex hydrides have been used in hydrogen storage systems. From complex hydrides, the hydrogen can be evolved (from the hydride) since interacting with water. The hydrolysis is irreversible, therefore, could not be utilized as host for rechargeable hydrogen storage systems. This class of materials are thermodynamically favorable, while kinetically plagued by high kinetics barriers to adsorption/desorption (Orimo et al. 2007).

As mentioned before, complex hydrides are potential hosts for solid-state hydrogen storage containing more than one metal nucleus. Sodium ($NaAlH_4$) and lithium ($LiAlH_4$) alanate, complex hydride (nano)composites (containing $LiAlH_4$, lithium amide ($LiNH_2$) and magnesium hydride (MgH_2) are examples of hydrogen storage complex hydrides.

Complex metal hydrides are a subclass of metal hydrides, consisting of M–H complexes in the crystal structure with the central ions from d- or p-block elements. These materials form often stoichiometrically, with a nonmetallic nature, where hydrogen ligands ordered at ambient conditions. Two broad families of complex metal hydrides are: (a) mononuclear complexes with only one transition metal atom, and one terminal hydrogen ligands, and (b) polynuclear complexes containing more than one metal nucleus. The mechanism of hydrogen storage can be proposed either physically, or chemically. The former is the simplest form of hydrogen bonding to the host material via physisorption. The second mechanism is the route of chemisorption, or strong hydrogen atomic bonding, to complex hydride materials. Complex hydrides require a high temperature for hydrogen release, as the hydrogen bonds are very strong. Crystallite size reduction of the host material (nanostructure), or by destabilizing the structure, the temperature of dehydrogenation can be reduced to allow reversible hydrogen storage. The anhydride complex hydride absorbs hydrogen to fully fill (charged) with hydrogen. Any active alteration in the surrounding conditions (pressure, voltage, temperature) can release hydrogen, in order to fully discharge (Fig. 4.4).

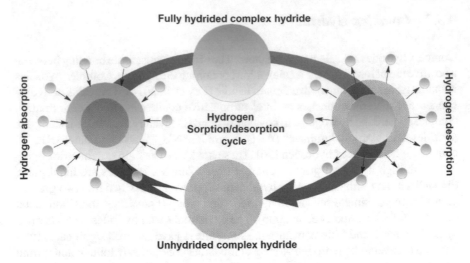

Fig. 4.4 Hydrogen sorption/desorption complex hydride cycle

4.3.3 Borohydrides

Boron is known as a polymeric cage (closo)-like compounds, and assorted into boro-hydrides, metallaboranes, and the carboranes (Massey 1964). Borohydrides itself can be classified into three classes called closo, nido, and arachno. Boron clusters of $[B_xH_x]^2$, $[B_xH_x]^4$, and $[B_xH_x]^6$ are adopted to closo (cage), nido (nest), and arachno (spider) structures (Atkins et al. 2006). In this class of materials with a general formula of B_xH_x, where $x = 5$–12, B_5H_5, B_6H_6, and $B_{12}H_{12}$ with trigonal-bi-pyramidal, octahedral, and icosahedral structures, respectively, are reported as effective textures for hydrogen storage systems (He et al. 2016). Borohydrides can adsorb/desorb hydrogen. Their hydrogen desorption occurs at elevated temperatures, typically at higher temperatures than their melting points (Nayak and Nayak 2016). The metal borohydrides can be melted at lower temperatures, which can further enhance the reactivity and kinetics of hydrogen desorption. Though these materials have high gravimetric and volumetric hydrogen content, their hydrogen desorption happen at high temperature ($T > 350$ °C). It is reported that nanoreduction in the particle size of the metal borohydrides can improve their reaction kinetics (Chen 2014). This is due to the reduction of diffusion distances, and an increase in the interfacial contact area of the reactants (Suchomel et al. 2018). Combining catalyst and borohydride in a storage system can enhance the storage performance of the system (Fig. 4.5). In this process, primarily, the H-atoms adsorbed on the Pd surface (H_{ad}) is the key species connecting the borohydride oxidation reaction (BOR), hydrogen evolution reaction (HER) and H absorption. In the anodic region, H absorption occurs by YH_2 (metal hydrides). The adsorbed hydrogen (H_{ad}) can be generated from $NaBH_4$ solutions, either from oxidation of BH_4^-, or from H_2O reduction (Chen et al. 2017).

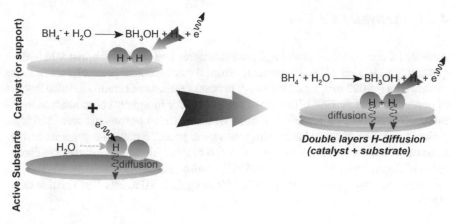

Fig. 4.5 Schematic representation of hydrogen storage hybrid system consisting a catalysis and an active substrate of metal hydride

4.4 Carbon-Based Materials

Reviews drawn from literature help us to draw a conclusion that the carbon materials are the best solid-state host for hydrogen storage (Alicia et al. 2017; Liu et al. 2011; Zhang et al. 2014a, b; Zhou et al. 2017). Carbon materials are a large group of nano-/microporous materials proposed for hydrogen storage owing to their high specific surface area. Activated carbons (ACs), carbon nanotubes (CNTs), multiwall carbon nanotubes (MWCNTs), graphene (GO), fullerene (F), and carbon dots (CDs) (Schlapbach and Züttel 2001). These materials can be subjected to the reversible hydrogen sorption according to the Langmuir isotherm model (Broom and Webb 2017). The wide range application of these materials has been restricted due to the low adsorption capacity at high cryogenic and low pressure (Froudakis 2011). Figure 4.6 shows an overall view of the process of hydrogen storage in carbonaceous materials.

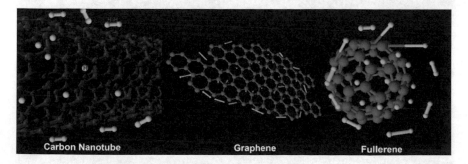

Fig. 4.6 Hydrogen storage carbonaceous materials

4.4.1 Activated Carbon

Activated Carbons (ACs) offer large pore structures and surface chemistry in hydrogen storage. ACs are attractive basically from the economical point of view, a large surface area ($\sim$2000 m^2/g), and their easy processabilities with controlled distribution of pores (Sircar et al. 1996). The use of organic and/or inorganic compounds as modifiers offers production activated carbons with controlled porous structure and high yield. Developed advanced technology allows to produce the homogeneous carbon adsorbents with benzene pore volume 0.3–0.6 cm^3/g (70–80%—volume of micropores), nitrogen surface area up to 1500 m^2/g, iodine adsorption capacity 40–70 wt% and methane adsorption capacity up to 160 mg/g (3.5 MPa, 293 K) (Vasiliev et al. 2007).

4.4.2 Fullerene

Fullerene plays as a nanocage. Fullerenes, due to their structural features, are efficient hydrogen storage hosts. Fullerenes are classified based on their number of structural carbons as C$_{70}$, C$_{76}$, C$_{80}$, C$_{84}$, etc., where C$_{60}$ is one of the most stable fullerenes (Malhotra et al. 2018) (Fig. 4.7). Fullerenes can be hydrogenated by several methods such as Birch reduction procedure, borane reduction, zinc–acid reduction, borane reduction, catalytic hydrogenation, and so on (Xiao et al. 2017).

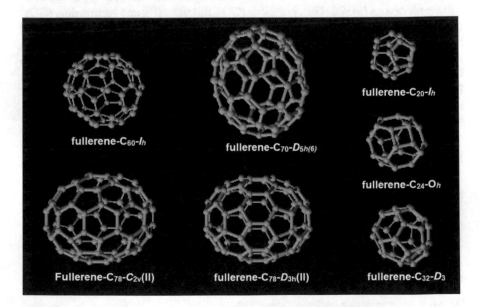

Fig. 4.7 Huge fullerene (C$_{60}$, C$_{70}$, C$_{78}$) and small fullerene (C$_{20}$, C$_{24}$, C$_{32}$)

Highly hydrogenated fullerenes are not stable, therefore it is potential to desorbed hydrogen after adsorption. Storing of hydrogen inside a fullerene cage, and inter-action between the hydrogen and the storage media are favored (Niaz et al. 2015). Theoretically, fullerenes have a maximum hydrogen storage capacity of 7.5 wt% at 0 K for C_{60}. This capacity can reach the proposed targets of future hydrogen storage systems, reported by US-DOE (Durbin et al. 2016).

4.4.3 Carbon Nanotube

Carbon nanotubes (CNTs) are novel nanocarriers, with a wide range of applications, either in academia or industries. CNTs can easily functionalize with certain chem-ical groups, organic, and inorganic materials. Functionalization of CNTs enhances their physical, chemical, and biological properties. CNTs, owing to their structural features, act as carriers for various molecules and atoms. The large surface area, physical dimensions, porosity, and various adsorbent/absorbent sites make CNTs suitable hosts for hydrogen storage (Elhissi and Dhanak 2018).

CNTs often refer to the single-wall (SWCNTs) or multiwall (MWCNTs). The position of the carbons governs the configuration and morphology of CNTs, for example, in SWCNTs there are zigzag, armchair, or chiral configurations, while in MWCNTs there are nanobuds, peapod, cup-stacked, etc. morphologies (Del-gado et al. 2008). Hydrogen adsorption binding energies are estimated based on the defects. The defects are governed by configurations; larger adsorption energies can be obtained for the configuration in which the hydrogen molecular axis perpen-dicular to the hexagonal carbon ring (Gayathri and Geetha 2007). Figure 4.8 shows various configuration of SWCNTs.

Inorganic functionalization of CNTs using metal (M), metal oxides (MOs), and mixed metal oxides (MMOs) have been received great attention in recent years. For instance, decoration of MWCNTs by TiO_2 shows hydrogen storage capacity of about

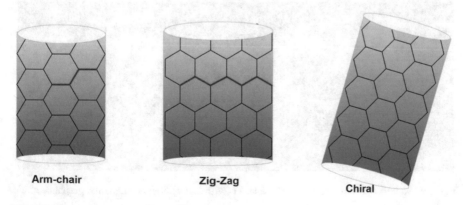

Arm-chair   Zig-Zag   Chiral

Fig. 4.8 Configurations of SWCNTs

2.02 wt% (Liu et al. 2011), CNTs/CoSe$_2$/MOFs nanocomposites illustrates superior hydrogen storage as compared to individual compartments (Zhou et al. 2016), cobalt (Co) decorated CNTs illustrates an enhanced hydrogen storage capacities of 717.3 mAh g^{-1} (2.62 wt% hydrogen) (Chang et al. 2014). All reviews drawn from the literature confirm the conclusion that the CNTs can be used as host for hydrogen storage.

4.4.4 Graphene

There is enormous interest in graphene-based nanostructures for energy storage, owing to its low weight, chemical stability, and low price. Theoretically, graphene is a large monolayer sheet constituted with several sp^2-bonded carbons. Based on its structure, there are two types of graphene; graphene with defects, and graphene without defects (Fig. 4.9). The structural features of graphene are the first assumption drawn in order to utilize graphene for various applications such as optical, electrical, mechanical, and electrochemical. The surface area of graphene is high enough, which makes it favorable for hydrogen energy storage. In addition, it is conductive which can functionalize like other carbonaceous materials with other molecules. In this class of materials, polygraphene is synthesized including two to nine layers of graphene sheets which restricted along a plane. This plane is called graphene nanoribbon. A multilayers graphene nanoribbon is called graphene nanofibers. Graphene oxide (GO) (Fig. 4.9) is the most chemically favored derivation of graphene (Pumera 2011). There are practical deficiencies in the utilization of graphene as a host in electrochemical energy storage devices. Some of these deficiencies are:

- Graphene has a much lower capacitance than the theoretical capacitance.
- The macroporous nature of graphene limits its volumetric energy density.
- Low packing density of graphene-based electrodes.

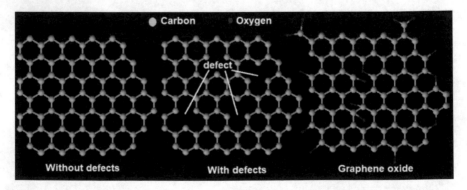

Fig. 4.9 Various structures of the graphene sheet (with and without defects), and graphene oxide (GO)

As mentioned in previous sections, microstructural properties—number of defects, stacking, the use of composite materials, conductivity, the specific surface area, the packing density, etc.—govern the electroactivity of carbon materials (here graphene). Therefore, the structural design of graphene is significant for electrode formation. Generally, six strategies (methodologies) are proposed for the optimal formation of electroactive graphene as follows:

- Non-stacking and 3D design concepts.
- Synthesis of highly packed graphene.
- Production of high performance/high structural graphene.
- Defects control.
- Functionalization with heteroatoms like oxygen, nitrogen, boron, and/or phosphorus.
- Formation of graphene (nano/micro) composites/hybrids.

4.5 Metal–Organic Frameworks (MOFs)

Metal–organic frameworks (MOFs) consist of multidentate organic ligands conjoining metal ion(s). These materials are known as the multi-hydrogen absorber due to their large surface areas, uniform porous structure, and the presence of metallic-core-redox species (Langmi et al. 2016). Three-dimensional frameworks of MOFs result in forming a network of channels. In this class of materials, narrow pores are more efficient than wide pores in the harvesting hydrogen uptake at room temperature. Structure-performance correlations in MOF systems can be divided into carboxylate-based frameworks, heterocyclic azolate-based frameworks, mixed-ligand/functionality systems, metal-cyanide frameworks, covalent organic frameworks (Murray et al. 2009). Figure 4.10 shows some important examples of MOF structures.

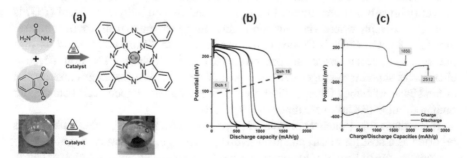

Fig. 4.10 **a** Production, **b** discharge capacity, and **c** charge–discharge efficiency of β-CuPc (Salehabadi et al. 2020)

This class of materials is extended mostly in three dimensions (3D), where the metal atoms are bridged by the organic ligand. Microporous MOFs with a large specific surface area (SSA) and pore volumes have been achieved a great interest in energy storage and conversion technologies. Their unique morphologies, functionalities, and relative ease of synthesis make MOFs potential in absorption technologies. Theoretically, it is also proved that MOFs are capable of adsorbed hydrogen. The shape selectivity and exact dimension of the pores are still big challenges. Unlike other porous substrates in hydrogen storage technologies, MOFs have several advantages together such as high crystallinity, porosity, surface area, hybrid nature, thermomechanical stability, gas and chemical stability, and decoration flexibility. MOFs can be decorated with metal and metal oxide (nano-)particles in order to enhance their practical hydrogen storage. Lithium (Li), Copper (Cu), Iron (Fe), Zinc (Zn), Nickel (Ni)—MOFs. Presence of aromatic frameworks can promote charge separation of charges, and making metal-doped more positive. This can lead to and providing strong stabilization to hydrogen (Niaz et al. 2015).

The model calculations were used in order to find the level of hydrogen adsorption and interactions with metal–organic frameworks (Mulder et al. 2008). It was observed that the high adsorption energy of MOF5 is comparable with nanostructured carbon materials. However, the relatively poor performance of MOF5 at room temperature is due to the weak interaction between H_2 and the framework surface. A number of large surface area MOFs with suitable gravimetric storage capacity has been synthesized (Gygi et al. 2016). Pb-MOF (Rahali et al. 2016), Cr-MOF (Ren et al. 2016), Rht-MOF (Suyetin 2017), Pd_9Ag_1-N-doped-MOF-C (Liu et al. 2017), Mg^{2+}/Ni^{2+}-MOF (Oh et al. 2017), α-Cu(II)-Phthalocyanine/$BaAl_2O_4$ (Salehabadi et al. 2017) are some important MOF systems, reported in hydrogen energy storage.

Copper phthalocyanine (CuPc), which is a MOF based sensitizer, has been recently examined in our laboratory for its electrochemical hydrogen storage performance (Salehabadi et al. 2020). It contains a planar framework with four iso-indole units. Nitrogen atoms in an azo position link these iso-indoles. These wide linkages lead to a longwave radiation absorption, probably with the higher $\pi \rightarrow \pi^*$ transition (Tedesco et al. 2016). CuPc can be synthesized in dry (without using solvent) or wet (with solvent) conditions. In a typical wet method, ethylene glycol (EG) is used as a solvent, where via continuous thermal treatments (from room temperature to ~185 °C), CuPc is formed (~163 nm width and 47 nm height) and used in preliminary expectations as an appropriate texture for hydrogen sorption. Electrochemically, superior hydrogen storage of around 1850 mAh/g discharge capacity and around 74% efficiency (charge–discharge) for rectangular nanocuboids beta copper phthalocyanine (β-CuPc) are reported (Fig. 4.10).

In another study (Salehabadi et al. 2017), a hybrid framework (CuPc/$BaAl_2O_4$) is reported as a host for electrochemical hydrogen storage. It is observed that the discharge capacity of $BaAl_2O_4$ nanoparticles has been enhanced upon addition of CuPc, from 900 mAh/g (3.2%) for $BaAl_2O_4$ with around 2.7 h discharge lost to 1500 mAh/g (~5.3%) with a discharge rate of 4.5 h for $BaAl_2O_4$/$BaCO_3$-CuPc nanohybrids. It is known that the presence of redox species like Cu(II) can enhance hydrogen storage capacities. Here, the overall reaction mechanism of hydrogen sorption is shown in Eq. 4.3:

$$BaAl_2O_4/BaCO_3\text{-}Cu^{(II)}Pc^{2-} + xH_2O \leftrightarrow BaAl_2O_4/BaCO_3\text{-}Cu^{(I)}Pc^{2-}H_x + xOH^-$$

$$(4.3)$$

Highly heterogeneous van der Waals potentials of MOFs within their pores govern the hydrogen molecules adsorption. The metal sites or metal-building units are preference of H_2 molecules to locate in particular loci. Weak adsorption sites created at room temperature, on the organic linkers, are further occupied after saturation of binding sites. The mechanism of hydrogen adsorption can be varied due to the system of MOFs. These systems are divided into the exposed metal sites, catenation/interpenetration, and spillover.

- **Exposed metal sites**—open metal coordination sites induced on the surface of metal–organic frameworks. Here, the H_2 can bind to metals in molecular systems, where metal–H_2 bond dissociation energies can reach very high value (Kaye and Long 2007).
- **Catenation** is interpenetration or interweaving of two or more identical and independent frameworks. Catenated MOF materials adsorb H_2 more strongly, with a high surface area, which correlated linearly with the overall H_2 uptake for homogeneous, physisorption-based systems (Yang et al. 2012).
- **Spillover**—In hydrogen spillover, molecular hydrogen dissociation occurs on a metal catalyst particle; a part remains attached to the catalyst and the rest diffuses on the support. The atomic hydrogen spillovers from the supported catalyst, therefore it can improve the contacts between the spillover source and the secondary receptor. This phenomenon can further enhance hydrogen capacity (Wang and Yang 2008).

4.6 Organic Polymers

Organic polymers have several strategic advantages over other materials in hydrogen storage technology. The organic polymers can undergo a wide variety of organic reactions. In addition, these materials prevented the inclusion of high molecular weight metals. Since the polymeric materials have a mechanical feature as compared to other materials in their fields, large surface area, and tendency to absorb water (especially proton conducting polymers), these materials can be a source for future hydrogen storage systems (Lopez et al. 2019). Cross-linked organic polymers and hypercrosslinked organic polymers are also developed for hydrogen storage systems. The highest specific surface area is reported for hypercrosslinked polystyrene. In addition, the polymers with intrinsic microporosity (PIMs) have been also devoted to hydrogen storage polymers, in which the pores are formed due to the inefficient packing of rigid polymer subunits (McKeown et al. 2006).

Hypercrosslinked polymers (also known as porous organic polymers) can reversibly absorb and release hydrogen via physisorption. These are covalently bonded (hydro) and thermally stable backbones exhibiting high and accessible surface areas, and properties, which are intriguing in the field of (Opto) electronics. N-heterocycle polymers fixed and stored hydrogen at atmospheric pressure through the formation of chemical bonds to form the corresponding alcohol and hydrogenated N-heterocycle polymers, respectively. Electrochemical hydrogenation using water as a hydrogen source was also effective, in which the polymer worked as a scaffold for hydrogenation. The hydrogenated polymers released hydrogen in the presence of catalysts at mild conditions. The potential of using organic polymers in the quest for finding new types of hydrogen-carrying and -storing materials that are very safe and portable is suggested.

Polymers can be crystalline, semi-crystalline, or amorphous. Highly crystalline polymers with 100% crystallinity do not exist, except in very low molecular weight polymers (single crystals). Crystalline polymers are rigid, where lacks the sorption sites (barrier effects) as well as the mobility of the chains to allow the mass transfer of gas molecules. Semi-crystalline polymers with more than one phase, i.e., crystalline and amorphous, are less packed with more gas sorption sites.

The amorphous polymers (or amorphous phase of semi-crystalline polymers) possess micro/nanopores. These polymers have advantages over other porous materials such as high porosity, metal-free, and lightweight, and easy to fabricate. Hypercrosslinked polystyrene is a macromolecule that can be prepared via simple Friedel–Crafts alkylation reactions method. The hypercrosslinked polymers can store hydrogen up to 5 wt% at high pressure and a low temperature.

Fused-ring polymers that are intrinsically microporous, are one class of organic with high surface areas ($\sim$1000 m^2 g^{-1}) and very efficient in hydrogen storage. These polymeric networks can reversibly adsorb hydrogen ($\sim$1.5 wt%) at low temperatures. Like MOFs, polymers can adsorbed–desorbed hydrogen, but minimally in ambient conditions. This limitation can be overcome by increasing the interaction energy of hydrogen with the polymer surface, for example, decreasing the pore size, which can enhance the adsorption enthalpy. This allows reversible hydrogen storage at room temperature (Kato and Nishide 2018).

Polyaniline and polypyrrole, alcohol-based polymers, 5,5',6,6'-tetrahydroxy-3,3,3',3'-tetramethyl-1,1'-spirobisindane copolymerized in bulk with tetrafluoroterephthalonitrile, hexachlorohexaazatrinaphthylene, and cyclotricatechylene are some examples of recently used polymers in hydrogen storage (Germain et al. 2006). Some important structures of hydrogen storage polymers are schematically presented in Fig. 4.11.

Two classes of organic–inorganic systems exist; class I has noncovalent interactions, while class II has some covalent interactions (Yang et al. 2009). These are novel classes of hydrogen storage systems. In the composite/hybrid materials, two or more components combine to overcome their weaknesses of each individual phase. It is observed that the hydrogen capacities in the composite materials reached values between those of the parental components. Therefore, a great attempt has been developed in order to fabricate novel materials with superior hydrogen

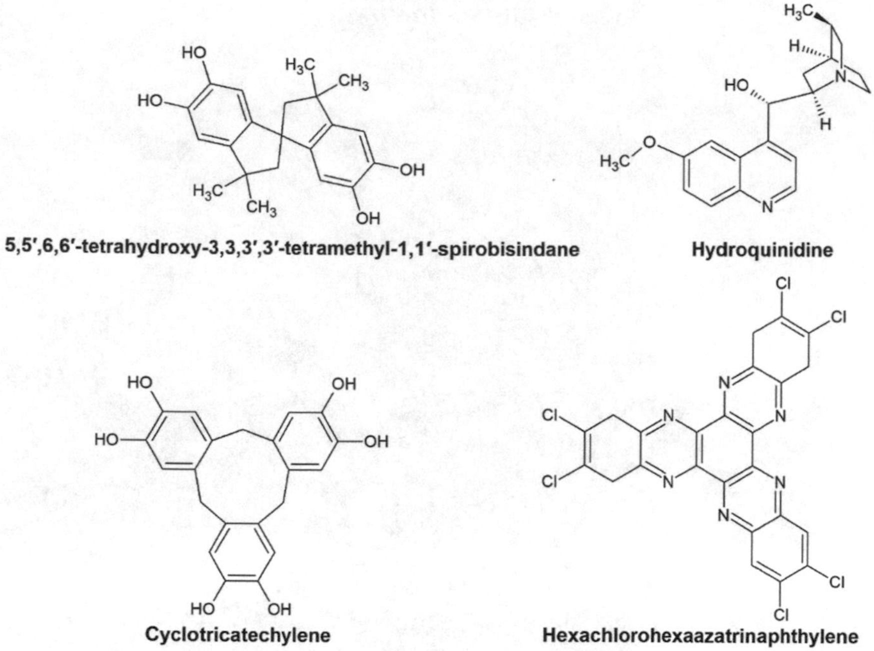
HO
HO
H_3C
CH_3
OH
H_3C
CH_3
OH
5,5′,6,6′-tetrahydroxy-3,3,3′,3′-tetramethyl-1,1′-spirobisindane
H_3C
H
HO
N
H
H_3C
O
N
Hydroquinidine
HO
OH
HO
OH
HO
OH
Cyclotricatechylene
Cl
Cl
N
N
Cl
N
Cl
N
N
N
Cl
Cl
Hexachlorohexaazatrinaphthylene

Fig. 4.11 Schematic representation of some polymeric-based hydrogen storage materials

storage performance. These materials involve in a large group of materials which are known as organic–inorganic materials (OIMs). The OIMs are classified into the covalent organic frameworks (COFs), microporous metal coordination materials (MMOMS), organotransition metal complexes (OTMCs), and polymeric-based composites (Aslanzadeh 2018).

A hydrogel of swelled polymers (water content of 25–35 wt%) with high concentration (1 M) of fluorenol/fluorenone units in the polymer or amorphous gel are used in order to investigate its potential in hydrogenation/dehydrogenation processes. It was observed that this unit was highly reactive to hydrogen, where the hydrogen exchanging reactions among the units were highly efficient. The units were fully hydrogenated, and then rapidly dehydrogenated throughout the polymer (Fig. 4.12).

In another study, fluorenone polymer composite/carbon fibers nanohybrids are used in order to study the hydrogen fixing/releasing in an aqueous medium. The hydrogen evolution of this conductive hybrid is obtained to be approximately in a similar trend to neat fluorenol polymer. The hybrid sheet is then hydrogenated applying −1.5 V (vs. Ag/AgCl).

Similarly, quinaldine (a N-heterocycle polymer) is known as a potential hydrogen carrier with a hydrogen capacity of 2.7 wt%. A composite containing poly (acrylic acid), quinaldine and 1,2,3,4-tetrahydroquinaldine is used as a host for hydrogen production/storage. In this admixture, hydroquinaldine polymer evolved hydrogen just by heating at around 80 °C in the presence of a catalyst solution (here iridium

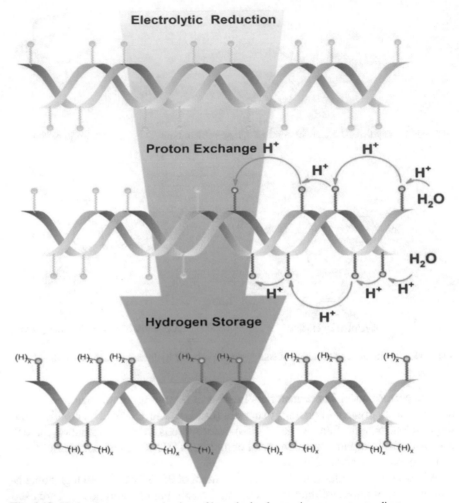

Fig. 4.12 Hydrogen storage mechanism of branched polymers in an aqueous medium

complex). A huge amount of hydrogen is evolved (160 ml per 1 g of polymer) at room temperature.

4.7 Clay and Zeolite

Clays and zeolites are widely studied as cheap, abundant, thermally stable, and non-toxic materials (Walker 2008). Owing to their weak interaction, clay and zeolite do not meet the volumetric storage targets, but the enthalpies of adsorption are significantly low, therefore, these materials can be used for hydrogen storage at ambient temperature and pressure (Weitkamp 2009). Zeolites are known as hydrogen trapping materials. In zeolites structure, the hydrogen can be trapped into the cavities of the molecular sieve at low temperature and released by raising the temperature.

Zeolites are aluminosilicates which are available in natural and synthetic forms. Physicochemical properties of zeolite make it appropriate for various applications such as adsorbents, molecular sieves, membranes, ion exchangers, energy storage, and catalysts in pollution control. It is known that the micropores in zeolite and its kinetic diameter are governed the hydrogen storage performance of zeolite. In addition, specific surface area and pore volume, the interaction of molecular hydrogen with the internal surfaces of the micropores, the stability of the molecular adducts, and the optimal storage temperature are the most important parameters which affect the hydrogen storage properties of zeolite. Based on the sodalite (cubic structure consists of an aluminosilicate cage network with Na^+) structural unit, various structures are proposed such as LTA (Zeolite A), FAU (X and Y zeolites), RHO, etc. Hydrogen storage capacity of zeolites highly depend on these structures and their respective cations. For example, the hydrogen contents in zeolite X is higher than that of zeolites A and rho. This phenomenon is due to the ability of large cations to occupy the 8-ring windows in these zeolites, blocking access of hydrogen to the zeolite cages, which is related to the surface area of the hosts. Figure 4.13 shows the relevance between hydrogen adsorption and BET surface area of various zeolites.

Natural clay is a porous material that is biocompatible, durable with approximately high hydrogen storage capacity. Montmorillonite (MMT) and halloysite nanotube (HNT) are two different classes of mineral clay with layered [tetrahedral-octahedral-tetrahedral (TOT)] and rounded [tetrahedral-octahedral (TO)] structures, respectively. The modified MMT and HNT are used in hydrogen storage systems, in which multistage hydrogen sorption are proposed. A template of montmorillonite K10 modified $Li_2CoMn_3O_8$ nanoparticles are studied by Salavati-Niasari research team (Ghiyasiyan-Arani and Salavati-Niasari 2018). They reported stable hydrogen storage of this composite. The electrochemical hydrogen storage performances of the samples are reported to be around 1040 mAh g^{-1} for pristine K10, and 1302 mAh g^{-1} for a composite containing 20 wt% $Li_2CoMn_3O_8$ nanoparticles.

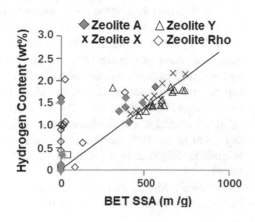

Fig. 4.13 Hydrogen adsorption and BET surface area of various zeolites (Anderson 2008)

In the next coming sections, the hydrogen storage properties of the MMOs will be discussed in terms of their chemistry, mechanism of hydrogen sorption, factors affecting hydrogen sorption, and hydrogen storage performances of various MMOs.

4.8 Mixed Metal Oxides (MMOs)

Metal oxides (MOs) and mixed metal oxides (MMOs) are most recent under studied materials in energy storage systems. Their wide application of MOs and MMOs are reported in gas storage, catalysis, and sensors (Gawande et al. 2012). However, few research outcomes in the physisorption of hydrogen in MMOs. It remains a difficult task of synthesizing suitable MMOs with appropriate physicochemical properties, which account for the strong interaction with hydrogen. In the conductive MOs and MMOs, the externally applied electric field could affect the dipole moment; therefore, the hydrogen could be more strongly adsorbed (Yuan et al. 2014). In addition, the presence of redox species with multi-oxidation states can be additionally enhanced the final hydrogen storage capacities.

A wide variety of mixed metal oxides (MMOs), containing alkali, alkaline, rare-earth and noble metals are investigated as active catalysts. They are also able to dissociate hydrogen molecules (H_2) to hydrogen atoms (H) and vice versa. This section summarizes the pertinent properties of the MMOs, covering their chemistry, mechanism, and efficiencies in hydrogen storage.

In general, MMO is defined as an oxide containing two or more metal cations, and sometimes known as complex metal oxides. The MMOs can be binary-, ternary-, quaternary-, etc. metal oxides owing to the presence of the number of metal cations. MMOs are either crystalline or amorphous, and are determined on their oxide compositions, for example, the oxide composition of crystalline MMOs can be formed perovskites, scheelites, spinels, palmeirites, and garnets possessed the general formula of ABO_3, ABO_4, AB_2O_4, $A_3B_2O_8$, $A_3B_5O_{12}$, respectively (Gawande et al. 2012). However, the exact arrangement of the metal cations differs from their coordination and the nature of the other cationic species present in the same structure (Cao et al. 2018). It remains unknown as to which of the metal cations constituted in the structure of MMOs, plays a role of active center.

MMOs play an important role in recent researches due to their structure, acid–base, and redox properties (Zeng et al. 2018). Many applications of MMOs are reported such as those in organic synthesis, green chemistry, fine chemical synthesis, and petroleum industries (Carrier et al. 2018). The prefix "nano" comes in many combinations throughout the MMOs (Aricò et al. 2005). Confinement effects of MMOs can lead to the fundamental manifestations of nanoscale phenomena as a key point of novel properties in various applications such as sensing (Guth and Wiemhöfer 2019), energy (Wang et al. 2018), and biological technologies (Aizawa 1994).

Nanoscale MMOs can be synthesized via two methods, either solution method (wet-chemistry) or vapor phase method (Carrier et al. 2018). In solution methods highly mobile reagents atomically mixed-up, where the incorporated molecules are crystallized in two stages, nucleation and growth. In the solution method, the reagents mix on atomic scale in a liquid medium; therefore, the reaction occurs in a fast diffusion and small diffusion distances. As a result, the overall reaction is carried out at approximately low temperatures. In the vapor phase method, super-saturation is required for successful nucleation to the point, where the solid materials are formed.

The terms "defects, "vacant sites", and "misplaced atoms" are features of MMOs (Wang et al. 2017). These disorders in the crystal formation of MMOs are thermodynamically favorable. Defects can influence the mechanical strength, electrical conductivity, and chemical reactivity of MMOs. Two type defects are considered as intrinsic and extrinsic defects; the former occurs in the pure substance, while the latter stems from the presence of impurities. The defects in MMOs are responsible for the enhancement of solid-state hydrogen storage. For example, it is reported that in a perovskite type MMOs, stable chemical bonds can be formed between proton and oxygen in the oxide, moreover, the vacancies in the perovskite could be replaced by protons. As a result, a substitutional OH^- ion defects can be formed (Deng et al. 2010).

Recently, MMOs have come to the fore of hydrogen storage technologies because of their chemical and physical properties, reversible capacity, and mechanical stability. Their superior hydrogen storage performance of MMOs takes part in several defects, redox reactions, and morphologies when drawing current.

The physical properties concepts of MMOs are associated with the structures and energy of ionic solids (MacChesney and Guggenheim 1969). It is critical to consider the nature and interaction between ions. The electronic structures of MMOs are also required to be propounded, based on electrons and ions interactions with each other. The study on the electronic structure of MMOs is important in order to understand the electric conductivity, magnetism, and many optical effects. The electronic activities of MMOs can be further reflected in hydrogen storage performances (Chen et al. 2018).

Semiconductors are electrically conductive. These are classified based on their composition. The band gap is the fingerprint of semiconductors. Generally, the semiconductors contain the p-block metals and group 13/14 metalloids. In semiconductors, the energies of the valence and conduction band separation is in the range of 0.2–4 eV (Atkins et al. 2006). The band gap in semiconducting materials can be influenced by particle size (Alivisatos 1996).

The conductivity of a typical semiconductor can be influenced by temperature. It is known that in the semiconducting MMOs, the hydrogen can be covalently bonded with surface atoms based on atomic configuration of the semiconductor. Curvy surface morphology of the semiconducting (nano)particles facilitate absorption/desorption of hydrogen (Cheng et al. 2014).

Theoretically, it is observed that the bulk surfaces of some semiconductors are metastable for high H-content (Chen 2014). Hydrogen is electrically active. The H-atoms (H^+ and H^-) seek out regions of high or low electronic charge density, where they interact with anions or cations, respectively. Hydrogen adsorption on sites of oxide model clusters (based metal oxides) is met the adsorption energy criteria with bond strengths ranging from 0.15 to 0.21 eV. In a MOs system (M $=$ Mg, Ba, Be), it is theoretically calculated that the energy profiles and kinetic constants for the splitting of the H_2 molecule reaction, is favored to adsorbed hydrogen on their sites, with a rapid uptake/release at operating temperatures and pressure (Gebhardt et al. 2014).

4.9 Summary

Renewable energy (RE), which is a part of primary energies, is a premier source in green energy technologies, but suffers from some drawbacks. Secondary energies like those that hydrogen is one of the solution to RE deficiencies, however, hydrogen suffers from its low density. Solid-state hydrogen storage technology is one of the solutions to all the above problems. Hydrogen storage materials can be used for onboard vehicle, material-handling equipment, and portable power applications. Carbon materials, MOFs, alloys, hydrides, MMOs, clay and zeolites, polymers, etc. are some examples of hydrogen storage materials.

Acknowledgements Fundamental Research Grant Scheme, Ministry of Education Malaysia 203/PTEKIND/6711274) and Universiti Sains Malaysia Postdoctoral Scheme.

References

M. Aizawa, Intelligent biomaterials, in *Advanced Materials '93* (Elsevier, New York, 1994), pp. 653–658. https://doi.org/10.1016/B978-1-4832-8380-7.50158-9

M. de Y. Alicia, Y. Oumellal, C. Zlotea, L. Vidal, M.G. Camelia, In-situ Pd–Pt nanoalloys growth in confined carbon spaces and their interactions with hydrogen. Nano-Struct. Nano-Objects **9**, 1–12 (2017). https://doi.org/10.1016/j.nanoso.2016.11.001

A.P. Alivisatos, Semiconductor clusters, nanocrystals, and quantum dots. Science (80-.) **271**, 933–937 (1996). https://doi.org/10.1126/science.271.5251.933

P.A. Anderson, Storage of hydrogen in zeolites, in *Solid-State Hydrogen Storage: Materials and Chemistry* (Elsevier Inc., New York, 2008), pp. 223–260. https://doi.org/10.1533/9781845694944.3.223

A.S. Aricò, P. Bruce, B. Scrosati, J.-M. Tarascon, W. van Schalkwijk, Nanostructured materials for advanced energy conversion and storage devices. Nat. Mater. **4**, 366–377 (2005). https://doi.org/10.1038/nmat1368

S.A. Aslanzadeh, Transition metal-metal oxide hybrids as versatile materials for hydrogen storage. Chin. J. Phys. **56**, 1917–1924 (2018). https://doi.org/10.1016/j.cjph.2018.09.011

P. Atkins, T. Overton, J. Rourke, M. Weller, F. Armstrong, *Inorganic Chemistry*, 4th ed. (Oxford University Press, Oxford, 2006)

S.J. Bonyhady, D. Collis, N. Holzmann, A.J. Edwards, R.O. Piltz, G. Frenking, A. Stasch, C. Jones, Anion stabilised hypercloso-hexaalane Al6H6. Nat. Commun. **9**, 3079 (2018). https://doi.org/10.1038/s41467-018-05504-x

D.P. Broom, C.J. Webb, Pitfalls in the characterisation of the hydrogen sorption properties of materials. Int. J. Hydrogen Energy **42**, 29320–29343 (2017). https://doi.org/10.1016/j.ijhydene.2017.10.028

C.-S. Cao, Y. Shi, H. Xu, B. Zhao, Metal–metal bonded compounds with uncommon low oxidation state. Coord. Chem. Rev. **365**, 122–144 (2018). https://doi.org/10.1016/J.CCR.2018.03.017

X. Carrier, S. Royer, E. Marceau, Synthesis of metal oxide catalysts, in *Metal Oxides in Heterogeneous Catalysis* (Elsevier, Amstredam, 2018), pp. 43–103. https://doi.org/10.1016/B978-0-12-811631-9.00002-8

C. Chang, P. Gao, D. Bao, L. Wang, Y. Wang, Y. Chen, X. Zhou, S. Sun, G. Li, P. Yang, Ball-milling preparation of one-dimensional Co–carbon nanotube and Co–carbon nanofiber core/shell nanocomposites with high electrochemical hydrogen storage ability. J. Power Sources **255**, 318–324 (2014). https://doi.org/10.1016/j.jpowsour.2014.01.034

J. Chen, Noble metal nanoparticle platform, in *Cancer Theranostics* (Elsevier, New York, 2014). https://doi.org/10.1016/B978-0-12-407722-5.00018-9

B. Chen, Z. Liu, C. Li, Y. Zhu, L. Fu, T. van Ree, Metal oxides for hydrogen storage, in *Metal Oxides in Energy Technologies* (Elsevier, New York, 2018), pp. 251–274. https://doi.org/10.1016/B978-0-12-811167-3.00009-2

J. Chen, J. Fu, K. Fu, R. Xiao, Y. Wu, X. Zheng, Z. Liu, J. Zheng, X. Li, Combining catalysis and hydrogen storage in direct borohydride fuel cells: towards more efficient energy utilization. J. Mater. Chem. A **5**, 14310–14318 (2017). https://doi.org/10.1039/C7TA01954H

H. Cheng, J. Wang, Y. Zhao, X. Han, Effect of phase composition, morphology, and specific surface area on the photocatalytic activity of TiO_2 nanomaterials. RSC Adv. **4**, 47031–47038 (2014). https://doi.org/10.1039/c4ra05509h

J.L. Delgado, M.Á. Herranz, N. Martín, The nano-forms of carbon. J. Mater. Chem. **18**, 1417–1426 (2008). https://doi.org/10.1039/B717218D

G. Deng, Y. Chen, M. Tao, C. Wu, X. Shen, H. Yang, M. Liu, Study of the electrochemical hydrogen storage properties of the proton-conductive perovskite-type oxide $LaCrO_3$ as negative electrode for Ni/MH batteries. Electrochim. Acta **55**, 884–886 (2010). https://doi.org/10.1016/j.electacta.2009.06.071

DOE Technical Targets for Hydrogen Storage Systems for Material Handling Equipment | Department of Energy (WWW Document), Dep. Energy, n.d. https://www.energy.gov/eere/fuelcells/doe-technical-targets-hydrogen-storage-systems-material-handling-equipment. Accessed 26 Nov 2018

DOE Technical Targets for Onboard Hydrogen Storage for Light-Duty Vehicles | Department of Energy (WWW Document). Dep. Energy. n.d. https://www.energy.gov/eere/fuelcells/doe-technical-targets-onboard-hydrogen-storage-light-duty-vehicles. Accessed 26 Nov 2018

D.J. Durbin, N.L. Allan, C. Malardier-Jugroot, Molecular hydrogen storage in fullerenes—a dispersion-corrected density functional theory study. Int. J. Hydrogen Energy **41**, 13116–13130 (2016). https://doi.org/10.1016/J.IJHYDENE.2016.05.001

A. Elhissi, V. Dhanak, Carbon nanotubes: applications in cancer therapy and drug delivery research, in *Emerging Nanotechnologies in Dentistry* (William Andrew Publishing, Boston, 2018), pp. 371–389. https://doi.org/10.1016/B978-0-12-812291-4.00018-2

G.E. Froudakis, Hydrogen storage in nanotubes & nanostructures. Mater. Today **14**, 324–328 (2011). https://doi.org/10.1016/S1369-7021(11)70162-6

M.B. Gawande, R.K. Pandey, R.V. Jayaram, Role of mixed metal oxides in catalysis science—versatile applications in organic synthesis. Catal. Sci. Technol. **2**, 1113–1125 (2012). https://doi.org/10.1039/c2cy00490a

V. Gayathri, R. Geetha, Hydrogen adsorption in defected carbon nanotubes. Adsorption **13**, 53–59 (2007). https://doi.org/10.1007/s10450-007-9002-z

J. Gebhardt, F. Viñes, P. Bleiziffer, W. Hieringer, A. Görling, Hydrogen storage on metal oxide model clusters using density-functional methods and reliable van der Waals corrections. Phys. Chem. Chem. Phys. **16**, 5382–5392 (2014). https://doi.org/10.1039/c3cp54704c

J. Germain, J. Hradil, J.M.J. Fréchet, F. Svec, High surface area nanoporous polymers for reversible hydrogen storage. Chem. Mater. **18**, 4430–4435 (2006). https://doi.org/10.1021/cm061186p

M. Ghiyasiyan-Arani, M. Salavati-Niasari, Effect of $Li_2CoMn_3O_8$ nanostructures synthesized by a combustion method on montmorillonite K10 as a potential hydrogen storage material. J. Phys. Chem. C **122**, 16498–16509 (2018). https://doi.org/10.1021/acs.jpcc.8b02617

U. Guth, H.-D. Wiemhöfer, Gas sensors based on oxygen ion conducting metal oxides, in *Gas Sensors Based on Conducting Metal Oxides* (Elsevier, New York, 2019), pp. 13–60. https://doi.org/10.1016/B978-0-12-811224-3.00002-0

D. Gygi, E.D. Bloch, J.A. Mason, M.R. Hudson, M.I. Gonzalez, R.L. Siegelman, T.A. Darwish, W.L. Queen, C.M. Brown, J.R. Long, Hydrogen storage in the expanded pore metal-organic frameworks M2(dobpdc) (M = Mg, Mn, Fe Co, Ni, Zn). Chem. Mater. **28**, 1128–1138 (2016). https://doi.org/10.1021/acs.chemmater.5b04538

S.S. Han, N.H. Goo, W.T. Jeong, K.S. Lee, Synthesis of composite metal hydride alloy of A2B and AB type by mechanical alloying. J. Power Sources **92**, 157–162 (2001). https://doi.org/10.1016/S0378-7753(00)00516-4

T. He, P. Pachfule, H. Wu, Q. Xu, P. Chen, Hydrogen carriers. Nat. Rev. Mater. **1**, 16059 (2016). https://doi.org/10.1038/natrevmats.2016.59

M. Hirscher, *Handbook of Hydrogen Storage: New Materials for Future Energy Storage* (Wiley, Chichester, 2010)

R. Kato, H. Nishide, Polymers for carrying and storing hydrogen. Polym. J. **50**, 77–82 (2018). https://doi.org/10.1038/pj.2017.70

S.S. Kaye, J.R. Long, Hydrogen adsorption in dehydrated variants of the cyano-bridged framework compounds $A_2Zn_3[Fe(CN)_6]2 \cdot xH_2O$ (A = H, Li, Na, K, Rb). Chem. Commun. **43**, 4486–4488 (2007). https://doi.org/10.1039/b709082j

H.W. Langmi, J. Ren, N.M. Musyoka, Metal-organic frameworks for hydrogen storage, in *Compendium of Hydrogen Energy* (Elsevier, 2016), pp. 163–188. https://doi.org/10.1016/B978-1-78242-362-1.00007-9

M. Latroche, Introduction to metal hydrides of AB3 compounds, in *Hydrogen Storage Materials. Advanced Materials and Technologies*, ed. by E. Burzo (Springer, Berlin, Heidelberg, 2018), p. 134. https://doi.org/10.1007/978-3-662-54261-3_21

M.B. Ley, L.H. Jepsen, Y.S. Lee, Y.W. Cho, J.M. Bellosta Von Colbe, M. Dornheim, M. Rokni, J.O. Jensen, M. Sloth, Y. Filinchuk, J.E. Jørgensen, F. Besenbacher, T.R. Jensen, Complex hydrides for hydrogen storage—new perspectives. Mater. Today **17**, 122–128 (2014). https://doi.org/10.1016/j.mattod.2014.02.013

E. Liu, J. Wang, J. Li, C. Shi, C. He, X. Du, N. Zhao, Enhanced electrochemical hydrogen storage capacity of multi-walled carbon nanotubes by TiO_2 decoration. Int. J. Hydrogen Energy **36**, 6739–6743 (2011). https://doi.org/10.1016/j.ijhydene.2011.02.128

Z. Liu, W. Dong, S. Cheng, S. Guo, N. Shang, S. Gao, C. Feng, C. Wang, Z. Wang, Pd_9Ag_1-N-doped-MOF-C: an efficient catalyst for catalytic transfer hydrogenation of nitro-compounds. Catal. Commun. (2017). https://doi.org/10.1016/j.catcom.2017.02.019

J. Lopez, D.G. Mackanic, Y. Cui, Z. Bao, *Designing polymers for advanced battery chemistries* (Rev. Mater., Nat, 2019). https://doi.org/10.1038/s41578-019-0103-6

J.B. MacChesney, H.J. Guggenheim, Growth and electrical properties of vanadium dioxide single crystals containing selected impurity ions. J. Phys. Chem. Solids **30**, 225–234 (1969). https://doi.org/10.1016/0022-3697(69)90303-5

A.J. Maeland, G.G. Libowitz, Hydrogen absorption in some A_2B intermetallic compounds with the $MoSi_2$-type structure ($C11_b$). J. Less Common Met. **74**, 295–300 (1980). https://doi.org/10.1016/0022-5088(80)90165-4

B.D. Malhotra, M.A. Ali, Functionalized carbon nanomaterials for biosensors, in *Nanomaterials for Biosensors* (William Andrew Publishing, Boston, 2018), pp. 75–103. https://doi.org/10.1016/B978-0-323-44923-6.00002-9

A.G. Massey, Boron. Sci. Am. **210**, 88–97 (1964). https://doi.org/10.1038/scientificamerican0164-88

N.B. McKeown, B. Gahnem, K.J. Msayib, P.M. Budd, C.E. Tattershall, K. Mahmood, S. Tan, D. Book, H.W. Langmi, A. Walton, Towards polymer-based hydrogen storage materials: engineering ultramicroporous cavities within polymers of intrinsic microporosity. Angew. Chem. Int. Ed. **45**, 1804–1807 (2006). https://doi.org/10.1002/anie.200504241

P. Millet, Hydrogen storage in hydride-forming materials, in *Advances in Hydrogen Production, Storage and Distribution*, ed. by A. Iulianelli, A. Basile (Woodhead Publishing, UK, 2014), pp. 368–409. https://doi.org/10.1533/9780857097736.3.368

R. Mohtadi, S. Orimo, The renaissance of hydrides as energy materials. Nat. Rev. Mater. **2**, 16091 (2017). https://doi.org/10.1038/natrevmats.2016.91

F.M. Mulder, T.J. Dingemans, H.G. Schimmel, A.J. Ramirez-Cuesta, G.J. Kearley, Hydrogen adsorption strength and sites in the metal organic framework MOF5: comparing experiment and model calculations. Chem. Phys. **351**, 72–76 (2008). https://doi.org/10.1016/j.chemphys.2008.03.034

L.J. Murray, M. Dinca, J.R. Long, Hydrogen storage in metal-organic frameworks. Chem. Soc. Rev. **38**, 1294–1314 (2009). https://doi.org/10.1039/b802256a

N.B. Nayak, B.B. Nayak, Temperature-mediated phase transformation, pore geometry and pore hysteresis transformation of borohydride derived in-born porous zirconium hydroxide nanopowders. Sci. Rep. **6**, 26404 (2016). https://doi.org/10.1038/srep26404

S. Niaz, T. Manzoor, A.H. Pandith, Hydrogen storage: materials, methods and perspectives. Renew. Sustain. Energy Rev. **50**, 457–469 (2015). https://doi.org/10.1016/j.rser.2015.05.011

H. Oh, S. Maurer, R. Balderas-Xicohtencatl, L. Arnold, O.V. Magdysyuk, G. Schütz, U. Müller, M. Hirscher, Efficient synthesis for large-scale production and characterization for hydrogen storage of ligand exchanged MOF-74/174/184-M (M = Mg^{2+}, Ni^{2+}). Int. J. Hydrogen Energy **42**, 1027–1035 (2017). https://doi.org/10.1016/j.ijhydene.2016.08.153

S. Orimo, Y. Nakamori, J.R. Eliseo, A. Züttel, C.M. Jensen, Complex hydrides for hydrogen storage. Chem. Rev. **107**, 4111–4132 (2007). https://doi.org/10.1021/CR0501846

V. Paul-Boncour, Introduction to hydrides of binary and pseudo-binary intermetallic compounds, in *Hydrogen Storage Materials, Advanced Materials and Technologies*, ed. by E. Burzo (Springer, Berlin, Heidelberg, 2018), p. 39. https://doi.org/10.1007/978-3-662-54261-3_9

M. Pumera, Graphene-based nanomaterials for energy storage. Energy Environ. Sci. **4**, 668–674 (2011). https://doi.org/10.1039/c0ee00295j

S. Rahali, M. Seydou, Y. Belhocine, F. Maurel, B. Tangour, First-principles investigation of hydrogen storage on lead(II)-based metal-organic framework. Int. J. Hydrogen Energy **41**, 2711–2719 (2016). https://doi.org/10.1016/j.ijhydene.2015.12.153

J. Ren, X. Dyosiba, N.M. Musyoka, H.W. Langmi, B.C. North, M. Mathe, M.S. Onyango, Green synthesis of chromium-based metal-organic framework (Cr-MOF) from waste polyethylene terephthalate (PET) bottles for hydrogen storage applications. Int. J. Hydrogen Energy **41**, 18141–18146 (2016). https://doi.org/10.1016/j.ijhydene.2016.08.040

A. Salehabadi, N. Morad, M.I. Ahmad, A study on electrochemical hydrogen storage performance of β-copper phthalocyanine rectangular nanocuboids. Renew. Energy **146**, 497–503 (2020). https://doi.org/10.1016/J.RENENE.2019.06.176

A. Salehabadi, M. Salavati-Niasari, T. Gholami, Effect of copper phthalocyanine (CuPc) on electrochemical hydrogen storage capacity of $BaAl_2O_4/BaCO_3$ nanoparticles. Int. J. Hydrogen Energy **42**, 15308–15318 (2017). https://doi.org/10.1016/j.ijhydene.2017.05.028

L. Schlapbach, A. Züttel, Hydrogen-storage materials for mobile applications. Nature **414**, 353–358 (2001). https://doi.org/10.1038/35104634

S. Section, *Hydrogen Storage, Multi-Year Research, Development, and Demonstration Plan* (2015)

S. Sircar, T.C. Golden, M.B. Rao, Activated carbon for gas separation and storage. Carbon N. Y. **34**, 1–12 (1996). https://doi.org/10.1016/0008-6223(95)00128-X

P. Suchomel, L. Kvitek, R. Prucek, A. Panacek, A. Halder, S. Vajda, R. Zboril, Simple size-controlled synthesis of Au nanoparticles and their size-dependent catalytic activity. Sci. Rep. **8**, 4589 (2018). https://doi.org/10.1038/s41598-018-22976-5

M. Suyetin, Rht-MOFs with triptycene-modified linkers for balanced gravimetric/volumetric hydrogen storage: a molecular simulation study. Int. J. Hydrogen Energy **42**, 3114–3121 (2017). https://doi.org/10.1016/j.ijhydene.2017.01.062

A.C. Tedesco, F.L. Primo, M. Beltrame, Phthalocyanines: synthesis, characterization and biological applications of photodynamic therapy (PDT), nanobiotechnology, magnetohyperthermia and photodiagnosis (theranostics), in *Reference Module in Materials Science and Materials Engineering* (Elsevier Ltd., 2016), pp. 1–6. https://doi.org/10.1016/B978-0-12-803581-8.02460-7

The Fuel Cell Technologies, Hydrogen Storage (WWW Document). Dep. Energy, n.d. https://www.energy.gov/eere/fuelcells/hydrogen-storage. Accessed 26 Nov 2018

L.L. Vasiliev, L.E. Kanonchik, A.G. Kulakov, D.A. Mishkinis, Activated carbon and hydrogen adsorption storage, in *Hydrogen Materials Science and Chemistry of Carbon Nanomaterials* (Springer, Dordrecht, 2007), pp. 633–651. https://doi.org/10.1007/978-1-4020-5514-0_80

G. Walker, *Solid State Hydrogen Storage: Materials and Chemistry*, 1st ed. (Woodhead Publishing, UK, 2008)

F. Wang, Y. Zhang, N. Yu, L. Fu, Y. Zhu, T. van Ree, Metal oxides in batteries, in *Metal Oxides in Energy Technologies* (Elsevier, 2018), pp. 127–167. https://doi.org/10.1016/B978-0-12-811167-3.00006-7

G. Wang, Y. Yang, D. Han, Y. Li, Oxygen defective metal oxides for energy conversion and storage. Nano Today **13**, 23–39 (2017). https://doi.org/10.1016/j.nantod.2017.02.009

L. Wang, R.T. Yang, New sorbents for hydrogen storage by hydrogen spillover—a review. Energy Environ. Sci. **1**, 268 (2008). https://doi.org/10.1039/b807957a

J. Weitkamp, Fuels—hydrogen storage—zeolites, in *Encyclopedia of Electrochemical Power Sources*, ed. by J. Garche (Elsevier, Netherland, 2009), pp. 497–503. https://doi.org/10.1016/B978-044452745-5.00330-0

S.-X. Xiao, C.-S. Huang, Y.-L. Li, Carbon materials, in *Modern Inorganic Synthetic Chemistry* (Elsevier, 2017), pp. 429–462. https://doi.org/10.1016/B978-0-444-63591-4.00016-1

S. Yang, X. Lin, W. Lewis, M. Suyetin, E. Bichoutskaia, J.E. Parker, C.C. Tang, D.R. Allan, P.J. Rizkallah, P. Hubberstey, N.R. Champness, K. Mark Thomas, A.J. Blake, M. Schröder, A partially interpenetrated metal–organic framework for selective hysteretic sorption of carbon dioxide. Nat. Mater. **11**, 710–716 (2012). https://doi.org/10.1038/nmat3343

S.J. Yang, J.Y. Choi, H.K. Chae, J.H. Cho, K.S. Nahm, C.R. Park, Preparation and enhanced hydrostability and hydrogen storage capacity of CNT@MOF-5 hybrid composite. Chem. Mater. **21**, 1893–1897 (2009). https://doi.org/10.1021/cm803502y

K. Young, D.F. Wong, L. Wang, J. Nei, T. Ouchi, S. Yasuoka, Temperature performance of AB5 hydrogen storage alloy for Ni-MH batteries. J. Power Sources **277**, 426–432 (2015). https://doi.org/10.1016/j.jpowsour.2014.10.093

C. Yuan, H.Bin Wu, Y. Xie, X.W.D. Lou, Mixed transition-metal oxides: design, synthesis, and energy-related applications. Angew. Chem. Int. Ed. **53**, 1488–1504 (2014). https://doi.org/10.1002/anie.201303971

L. Zeng, Z. Cheng, J.A. Fan, L.-S. Fan, J. Gong, Metal oxide redox chemistry for chemical looping processes. Nat. Rev. Chem. **2**, 349–364 (2018). https://doi.org/10.1038/s41570-018-0046-2

C. Zhang, J. Li, C. Shi, C. He, E. Liu, N. Zhao, Effect of Ni, Fe and Fe-Ni alloy catalysts on the synthesis of metal contained carbon nano-onions and studies of their electrochemical hydrogen storage properties. J. Energy Chem. **23**, 324–330 (2014a). https://doi.org/10.1016/S2095-4956(14)60154-6

Y. Zhang, H. Ren, Y. Cai, T. Yang, G. Zhang, D. Zhao, Structures and electrochemical hydrogen storage performance of Si added A2B7-type alloy electrodes. Trans. Nonferrous Met. Soc. China **24**, 406–414 (2014b). https://doi.org/10.1016/S1003-6326(14)63076-4

Y. Zhang, W. Zhang, X. Song, P. Zhang, Y. Zhu, Y. Qi, Effects of spinning rate on structures and electrochemical hydrogen storage performances of RE–Mg–Ni–Mn-based AB2-type alloys. Trans. Nonferrous Met. Soc. China **26**, 3219–3231 (2016). https://doi.org/10.1016/S1003-6326(16)64454-0

W. Zhou, J. Lu, K. Zhou, L. Yang, Y. Ke, Z. Tang, S. Chen, $CoSe_2$ nanoparticles embedded defective carbon nanotubes derived from MOFs as efficient electrocatalyst for hydrogen evolution reaction. Nano Energy (2016). https://doi.org/10.1016/j.nanoen.2016.08.040

H. Zhou, L. Zhang, S. Gao, H. Liu, L. Xu, X. Wang, M. Yan, Hydrogen storage properties of activated carbon confined $LiBH_4$ doped with CeF_3 as catalyst. Int. J. Hydrogen Energy **42**, 23010–23017 (2017). https://doi.org/10.1016/J.IJHYDENE.2017.06.193

Chapter 5
Essential Parameters Identification of Hydrogen Storage Materials

Abstract In this chapter, a brief description of the requirements of a hydrogen storage system is given. The weak interaction of hydrogen within pores (sites) needs to be understood in order to design and develop porous materials for hydrogen sorption. The measurements are based on the amount of hydrogen adsorbed as a function of pressure, temperature, the enthalpies of adsorption, and the adsorption/desorption characteristics (Thomas in Hydrogen adsorption and storage on porous materials. Catal. Today 120.389–398, 2007).

5.1 Introduction

Research on hydrogen storage materials, their production routes, and real applications are of great value (Salehabadi et al. 2020). Computational studies and experimental observations can be understood when a real hydrogen storage system build-up. The demands for hydrogen vehicles have increased due to environmental concern and escalating fossil fuels price. It is reasonably feasible if the hydrogen can store on solid hosts, which is more reliable, cheaper, and safer (Salehabadi et al. 2019). As mentioned in previous chapters, hydrogen storage materials are hosts for chemical or physical adsorption/absorption of hydrogen and should possess appropriate thermodynamic and kinetic properties. The thermodynamics are important in the hydrogenation–dehydrogenation or adsorption–desorption processes (Song et al. 2012). In general, some important characteristics (let call MUST) are expected for an adoptable solid-state hydrogen storage system (Viswanathan 2017);

I.   The overall view of the system MUST be thermodynamically favorable, i.e., not much heat requirement.
II. The adsorption and desorption MUST be kinetically favorable, i.e., fast kinetic.
III. The system MUST be reversible, i.e., recoverable or returnable to its baseline easily, just after exposure.
IV. The extent of storage MUST possesses approximately a large volumetric and gravimetric density, i.e., adaptable for mobile applications.
V.   The hosting materials MUST withstand a sufficient cycle number.

VI. The materials used MUST be favorable in terms of heat transfer medium, durability, and mechanical strength.

In the following sections, the above major characteristics will be discussed.

5.2 Thermodynamics of Hydrogen Storage

In hydrogen storage technology, hydrogen after generation should be stored on solid-state materials, however, some thermodynamic obstacles must be solved, in order to achieve high efficiency. As we discussed before, the solid-state hydrogen adsorption is followed by hydrogen desorption, which occurs through the process of two bonds breaking; (a) hydrogen molecules dissociation ($H_2 \rightarrow H + H$), and (b) between hydrogen with some other substances ($MH \rightarrow M + H$). Therefore, the thermodynamic of hydrogen storage must be improved.

Pressure-composition isotherms (PCI) (based on Van't Hoff theory Eq. 5.1) are used to ascribe the thermodynamic aspects of hydride formation.

$$\ln\left(\frac{p_{eq.}}{p^\circ_{eq.}}\right) = -\frac{\Delta H}{R \times T} + \frac{\Delta S}{R} \tag{5.1}$$

In this isotherm (Fig. 5.1), the amount of stored hydrogen can be determined from the length of the plateau pressure (equilibrium), where solid solution and hydride phases coexist. Here, the plateau pressure ($P_{eq.}$) depends on temperature (T), changes of enthalpy (ΔH) and entropy (ΔS). In this isotherm, there are two phases, α (solid solution) and β (hydride phase). In the β-phase, the hydrogen pressure and concentration are directly respective to each other. These two phases end at a critical point (T_c). Any transitions above T_c—from α- to β-phase—are continuous. From the PCI

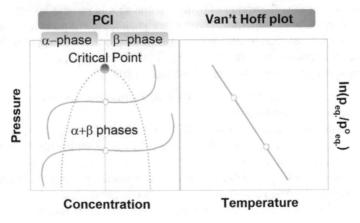

Fig. 5.1 Schematic representation of pressure-composition isotherm (PCI), and Van't Hoff plots

(Fig. 5.1), it can be also deduced that upon increasing hydrogen pressure the amount of absorbed hydrogen will increase.

5.3 Adsorption/Desorption Kinetic

Hydrogen adsorption/desorption kinetics must be pointed out as important as the hydrogen adsorption/desorption thermodynamic, especially in electrochemical hydrogen storage technology, due to the need for fast charging and discharging of hydrogen. Most of the solid-state hydrogen storage materials exhibit unsatisfactory charge and discharge kinetics. Surface modification and spillover mechanism (which can be observed in layered materials such as carbon nanotube) are used in order to improve the kinetics of hydrogen storage. Thomas (2007), in a study on kinetics of hydrogen storage on porous materials, expressed that a fast adsorption/desorption kinetics and relatively small adsorption enthalpies are required for hydrogen adsorption on many porous materials. He indicated that the physisorption on porous materials governed the fast recharging with hydrogen. In addition, Thomas reported the kinetic trapping of hydrogen in MOF materials, which can affect the temperature dependence of hydrogen adsorption.

Theoretically, in porous materials like MOFs, the adsorption kinetics can be measured using diffusion model established by Saha et al. (2008). They proposed a spherical coordinate at which the heat transfer between particle and surrounding fluid is neglectable, as (Eq. 5.2),

$$\frac{\partial q}{\partial t} = \frac{1}{r^2} \times \frac{\partial}{\partial r}\left(r^2 \times D_c \times \frac{\partial q}{\partial r}\right) \tag{5.2}$$

r: radius of the equivalent sphere
D_c: intracrystalline diffusivity
q: adsorbed hydrogen amount (wt%)
r: radial position
t: time.

For constant diffusivity, the average adsorbate concentration in the particles, and fractional uptake $(m/m_{max}) > 70\%$, the following equation (Eq. 5.3) can be obtained from Eq. 5.2,

$$1 - \frac{m_t}{m_{max}} = \frac{6}{\pi^2} \exp\left\{\frac{-\pi^2 \times D_c \times t}{r_c^2}\right\} \tag{5.3}$$

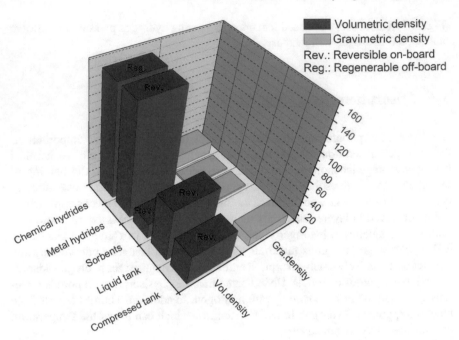

Fig. 5.2 A comparison between hydrogen storage technologies, reversibility, volumetric, and gravimetric densities

5.4 Reversibility

Reversibility is the ability of a device/material to recover/return to its baseline condition, after exposure. Reversibility is one of the most important criterion in a hydrogen storage system. It is quite clear that solid-state materials (except carbon materials which require high temperature to release hydrogen) in physical hydrogen adsorption have an approximately large reversible hydrogen storage capacity (Wang et al. 2005). Basically, among the methods for hydrogen storage, some of the methods with a high volumetric and gravimetric density have reversible hydrogen storage. According to the U.S. Department of Energy (DOE), the physical-based hydrogen storage and material-based hydrogen storage are categorized as either "reversible onboard" or "regenerable off-board" (Hwang and Varma 2014). Figure 5.2 compares the methods of reversible hydrogen storage with high volumetric and gravimetric densities.

5.5 Experimental Setup

It is crucial to measure accurate hydrogen adsorption on solid-state materials, while early work on hydrogen adsorption on materials did not validate properly, in terms of information and measurements. Gravimetric and volumetric measurements provide direct and indirect measurements of adsorption, respectively. The following basic system requirements are needed with gravimetric isotherm and isobar measurements, in order to obtain occur results (Thomas 2007):

- An ultra-clean high vacuum system with appropriate turbopumps.
- A hydrogen purification system add-on.

As a subgroup of the above technologies, chronopotentiometry has been achieved a great interest in solid-state electrochemical hydrogen storage systems. Chronopotentiometry is an electrochemical technique in which a constant current is caused to flow between two electrodes; the potential of one electrode is monitored as a function of time with respect to a suitable reference electrode. The system setup is made by three electrodes;

- **Working electrode (WE)**: The working electrode is the electrode in an electrochemical system where the reaction of interest is occurring, in conjunction with a counter electrode, and a reference electrode. WE materials provide good electron transfer properties towards the substrate. The WE materials exhibit high activation energy for electron transfer in the principal competing reaction. For example, the most significant competing reactions in the presence of water are an evolution of oxygen at the anode and hydrogen at the cathode.
- **Counter electrode (CE)**: The counter electrode is an electrode in an electrochemical system where the current circuit is closed. The based materials for CE are inert materials like platinum (Pt), gold (Au), graphite, glassy carbon (GCE). This electrode is not a part of electrochemical reaction profile.
- **Reference electrode (RE)**: RE provides a standard for the electrochemical measurements, in which stable and accurate results can be obtained in the measurement circuit. In addition, RE provides a stable reference potential, especially in the pH-sensitive systems.

The electrochemical reaction occurs on WE. In a typical hydrogen storage system, WE are a sample electrode, where the sample coated on a surface of metallic sheets of redox inert materials (like glassy carbon, copper, nickel, etc.), either by conventional solvent casting or mechanical deposition. The applied potential of the WE counts as a function of the RE.

A RE is used as a reference point, where the potential of other electrodes can be counted. The most commonly used RE in aqueous media are saturated calomel electrode (SCE), standard hydrogen electrode (SHE), and the AgCl/Ag electrode.

The CE is used to complete the electrical circuit. In three electrochemical setups, as potential is applied to the WE, a reduction or oxidation of the analyte can occur, where current flows. In this setup, a current between the WE and CE records as electrons flow. An inert wire/disk is typically used as a CE.

In this electrochemical setup, the primary outcome of the recorder such as discharge capacity (DC), the hydrogen content (HC), the hydrogen density (HD), and the charge–discharge hydrogen efficiency ($H_{eff.}$) of the systems need to calculate. The DC is the summation of all successive time (t), and mass (m) associated with each small discharge at the operating current (i) Eq. 5.4;

$$DC_{max} = (i/m) \times \Sigma t \tag{5.4}$$

This factor is the first parameter that should be considered, which can further reflect in the hydrogen storage efficiency of the system. The HC is the second important factor, which can be calculated based on discharge capacity of the system. It is known that 1 m^3 of water ($H_2O \rightarrow 2H^+ + OH^-$) contains around 111 kg of hydrogen (1 m^3 = 111 kg). On the other hand, 1 m^3 of hydrogen is around 71 kg of hydrogen (1 m^3 = 71 kg). Therefore, the hydrogen content of the adsorbent (here MMOs) in 1 mA can be calculated as Eq. 5.5;

$$HC = (3.55 \times DC)/1000 \tag{5.5}$$

The efficiency of a battery can be calculated as the amount of power discharged by the battery divided by the amount of power delivered to the battery as Eq. 5.6;

$$\%Efficiency = (Discharge/Charge) \times 100 \tag{5.6}$$

This considers the loss of energy to heat, which warms up the battery.

The cycling stability of the alloy is characterized by the capacity conservation rate, which is defined as the ratio of discharge capacity after stabilization by the maximum discharge capacity, Eq. 5.7;

$$S = 100 \times C_s/C_{max} \tag{5.7}$$

where C_{max} is the maximum discharge capacity (mAh/g) and C_s is the discharge capacity after stabilization (mAh/g).

5.6   Composition and Formulation Ingredients

Hydrogen storage materials especially alloys, hydrides, MMOs, and MOFs are generally synthesized utilizing various methods. An improvement of the storage performance of these materials can be achieved by selecting the appropriate elements in the formulation ingredients. In this section, an attempt has been developed to classify the solid-state hydrogen storage materials based on their abundance of central ions.

5.6.1 Aluminum (^{13}Al)

Aluminum is one of the most abundant metal ions in the structure of novel hydrogen storage materials like MMOs. It has practically utilized in battery owing to its appropriate theoretical voltage and specific energy (Li and Bjerrum 2002). Nanoscales $ZnAl_2O_4$ (Gholami et al. 2018a), $BaAl_2O_4$ (Salehabadi et al. 2017b), $NiAl_2O_4$ (Gholami et al. 2018b), $CoAl_2O_4$ (Gholami et al. 2016), $Sr_3Al_2O_6$ (Salehabadi et al. 2018b), $Dy_3Al_2(AlO_4)_3$ (Salehabadi et al. 2018d) are some of the reported MMOs in a typical hydrogen storage system. These MMO nanoparticles are determined on their oxide compositions to be in spinel (AB_2O_4), poly-crystal ($A_xB_yO_z$), and garnet ($A_3B_5O_{12}$) forms. Considerably high discharge capacity and hydrogen content of the above materials are reported. In this level of MMOs, the complex structures like polycrystals (Salehabadi et al. 2018b) and garnets (Salehabadi et al. 2018d) show superior discharge capacities as compared to other M-Al-Os. Interestingly, the composite/hybrid materials exhibit higher value of discharge capacities and respective hydrogen contents, due to their larger specific surface area (Gholami et al. 2018b), presence of redox species (Gholami and Salavati-Niasari 2016), or/and multilevel hydrogen storage possibilities (Salehabadi et al. 2017a). In a study on the effect of copper phthalocyanine (CuPc) sensitizers on the hydrogen storage performance of spinel $BaAl_2O_4$ nanoparticles, it is reported that the impregnation of host by CuPc, can positively affect the discharge capacity. A 1500 mAh/g of discharge capacitance are obtained for $BaAl_2O_4/BaCO_3$-CuPc nanocomposites after 15 cycles, higher than that of pristine $BaAl_2O_4/BaCO_3$ (900 mAh/g). It is expected that the obvious value of discharge capacity is laid down between the discharge capacity of host nanoparticles, and active modifier [here CuPc with ~2500 mAh/g discharge capacity (Salehabadi and Salavati-Niasari 2018)]. Here, it is proposed that there are interactions between the active sites of CuPc and hydroxyl groups on the surface of $BaAl_2O_4$, in the form of H-bonding. These two layers systems, i.e., $BaAl_2O_4$ and CuPc can both adsorb hydrogens. In addition, CuPc is a metal complex with a central active redox species (Cu^{2+}), which can easily reduce following the reaction, Eq. 5.8;

$$Cu(II)Pc^{2-} + xH_2O + xe^- \leftrightarrow Cu(I)Pc^{2-}H_x + xOH^-. \tag{5.8}$$

As we will discuss later, the presence of redox species can boost the hydrogen storage performances.

5.6.2 Iron (^{26}Fe)

Iron is the second interesting ions in energy storage systems. It has long been considered as an attractive metal ion for a wide variety of energy applications. Magnetic collapse in Iron (Fe) containing MMOs is predicted from band widening (Cohen et al. 1997). Theoretically, it is known that the Fe in a typical ABO_x formulation is high-spin in the A site and low-spin in the B site. The presence of Fe in the B site can affect the magnetic moment. Generally, the B site is smaller than the A site, therefore, the bandwidths are larger as Fe placed in the B site, therefore, a variation in magnetic behavior can take place.

Iron-based MMOs are used as anodes in batteries that form fewer dendrites and rechargeable in alkaline aqueous solutions. A typical energy storage system with M-Fe-O_x electrode and alkaline electrolyte can be favorable since their appropriate theoretical capacity, safety, elemental abundance, and low cost. $La_{0.4}Sr_{0.6}FeO_3$ (Deng et al. 2009), $LaFeO_3$ (Deng et al. 2010), $Dy_3Fe_5O_{12}$ (Salehabadi et al. 2018c), and MgH_2 containing $SrFe_{12}O_{19}$ (Mustafa et al. 2016), $NiFe_2O_4$ (Wan et al. 2015), $CuFe_2O_4$ (Ismail et al. 2018), $MnFe_2O_4$ (Zhai et al. 2012) nanocomposites are reported as potential hydrogen storage MMOs.

Huo et al. (2016) proposed a mechanism in order to describe the hydrogen storage of α-Fe_2O_3 particles in an aqueous solution. They expressed that the H^+ can transfer in the MO lattice most probably due to the reaction of H^+ in the interior of the Fe_2O_3 particles. They confirmed this expectation using the first-principle simulation to show the covalent bonding between H^+ and O in the interior of the Fe_2O_3. The same expectation has been considered to conclude the high discharge capacities of $DyFeO_3$ and $Dy_3Fe_5O_{12}$ nanoparticles (Ali Salehabadi et al. 2018c). In this study, a facial combustion method is used for the synthesis of perovskite and garnet Dy-Fe-O nanoparticles. The physicochemical properties of these to crystalline phases rely on the electrochemical properties, where, an enhanced cyclic voltammogram of $DyFeO_3$ containing minor phases of Fe_2O_3 is reported as compared to $Dy_3Fe_5O_{12}$ nanoparticles. This trend is further reflected in hydrogen storage performance of the sample. It is highlighted that the superior discharge capacity of $DyFeO_3$, in addition to the presence of redox species in its structure, can be also described since minor Fe_2O_3 phase existed.

Perovskite $LaFeO_3$ powders are also related by Deng et al. (2010), were synthesized using a stearic acid combustion method as negative electrodes for nickel/metal hydride (Ni/MH) batteries. They expressed that the discharge capacity of this electrode is enhanced upon increasing temperature. For example, the discharge capacities during the first three cycles stayed steady at about 80 mAh/g, 160 mAh/g, and 350 mAh/g at 298 K, 313 K, and 333 K, respectively.

The composite/hybrid materials containing M-Fe-O are investigated in order to enhance the hydrogen storage performance of the host texture. In a study of hydrogen storage properties of layered-structure-based nanocomposites like multiwall carbon nanotubes (MWCNTs) (Salehabadi et al. 2018a), it is reported that superior properties are obtained, compared to the pristine host materials. In this nanocomposite (MWCNTs/$Dy_3Fe_5O_{12}$), a stable hydrogen storage capacity of 1000 mAh/g is recorded, higher than MWCNTs itself ($\sim$420 mAh/g). The mechanism of hydrogen sorption in this system relied not only on their ability to surface adsorption of $Dy_3Fe_5O_{12}$, but also on the intrinsic and unique features of MWCNT spillover. In spillover, the MMO can act as an activated center which catalyzes hydrogen molecule dissociation (Strobel et al. 2006), where the hydrogen atoms spillover from the additive sites to the carbon network/layers, and bounded (Pyle et al. 2016). The hydrogen storage in M-Fe-O materials is a step-wise electrochemical process in which the hydrogen atoms may be intercalated into the oxide lattice by the formation of a homogeneous solid solution (Esaka et al. 2004).

5.6.3 Vanadium (^{23}V)

Vanadium is used as a central ion in transition metal vanadium oxides (TMVOs) for the production of rechargeable batteries. Transport properties in cation exchange membranes (Wu et al. 2018), electrical and optical properties (MacChesney and Guggenheim 1969; Verleur et al. 1968), and energy storage/production (Saïdi et al. 2002) properties of vanadium ions in various structural systems have been reported. Vanadium (V) has five valence electrons, therefore it can adopt multiple oxidation states. This variation in oxidation states can be further reflected in the properties of MMOs containing V ion (Haber et al. 1986).

The electrochemical performance of these ions is involved in multistep vanadium reduction/oxidation reactions. ZnV_2O_4 is used as a host in aqueous rechargeable zinc-ion batteries (ARZIBs), with respect to its unique structure and properties. Layered α-$Zn_2V_2O_7$ is highlighted as a safe, and low-cost materials that exhibited excellent capacity retention of 85% after 1000 cycles at an ultra-high current drain of 4000 mA/g (Sambandam et al. 2018).

ZnV_2O_4 glomerulus nano/microspheres are reported to expose its hydrogen storage potential (Butt et al. 2014). In this study, the hydrogen adsorption/desorption profiles of this material are investigated based on its structural parameters. The authors proposed that any changes in structural parameters can occur after hydrogen storage, therefore, using Reitveld analysis they refined lattice parameters to be 12.5% less than pre-measured results of hydrogen storage. They finally concluded that the lattice planes are more oriented in the sample after hydrogen adsorption, i.e., reducing defects and vacancies in nanostructures. On the other hand, they quantitatively found a correlation between the physical properties of ZnV_2O_4 nanosystems and hydrogen absorption.

The hydrogen storage performance of the $Mg_{17}Al_{12}/V_2O_5$ nanocomposites is investigated by Wu et al. (2018). In this research, they measured the hydrogenation/dehydrogenation temperature and the reversible hydrogen storage properties of the $Mg_{17}Al_{12}$ alloys, and $Mg_{17}Al_{12}/V_2O_5$ nanocomposites and expressed that upon increasing V_2O_5, the hydrogen properties of the working system gradually improved.

Hereupon, it can be summarized that vanadium involved in MMOs can significantly affect the electrochemical properties of the product. In addition, the combination of M-V-Os or V-Os with any active substrates can be enhanced hydrogen storage performances since multilevel hydrogen sorption can occur. It seems that graphene, CNT, polymers, and clay can be good choice in these criteria owing to their conductivity, stability, porosity, and reactivity.

5.6.4   Cobalt (^{27}Co)

Cobalt in hydrogen storage materials containing trivalent cobalt is justified by researchers owing to its range of properties like electrochemical reactivity, magnetic and magneto-optical, dielectric, etc. for low-cost and environmentally friendly energy storage/conversion technologies. The wide range applications of M-Co-O can be found in lithium-ion batteries, electrochemical capacitors, metal-air batteries, and fuel cells. It is reported that in MMOs containing cobalt ions species, the morphologies, sizes, compositions, and micro/nanostructures governed their applications as electrode materials (Yuan et al. 2014). However, further development requires next-generation hydrogen storage systems.

Electrochemical hydrogen storage properties of $Ba_2Co_9O_{14}$ nanoparticles are reported as having 800 mAh/g discharge capacity (Razavi et al. 2019). In this research article, the $Ba_2Co_9O_{14}$ nanoparticles are synthesized via a Pechini method in the presence of ethylene glycol (EG) and poly(ethylene glycol)-600 (PEG-600) consisting the particles ranging from 10 to 30 nm. The authors expressed that the ultimate hydrogen storage performance of this MMO can be due to its physicochemical properties. In addition, it is mentioned that the hydrogen sorption mechanism occurred not only by physisorption among a series of bonding between protons and oxygen, but also redox reactions (Ce^{IV}/Ce^{III}, Ni^{III}/Ni^{II}, Co^{III}/Co^{II}).

The majority of the articles in the cobalt-containing MOs/MMOs are focused on the composite materials of cobalt oxides in various compositions such as Co_3O_4/N-RGO (Li et al. 2018), and $Ti_{1.4}V_{0.6}Ni$ @ C @ Co_3O_4 (Lin et al. 2016). In the former composition (Co_3O_4/N-RGO), the synergistic effect between Co_3O_4 nanocubes and modified graphene (RGO) is described on the major obstacle of its electrochemical reaction during charging/discharging, where a high capacity of around 223 mAh/g at a specific current of 3000 mA/g is reported. This capacity is 3.2 times higher compared to the available commercial alloy electrode with 68.7 mAh/g discharge capacity, which originated from the unique structure and 3D porous morphology of the composites. The features of this nanocomposite for electrochemical energy storage are significant since the RGO (which is known as conductive substrate) and Co_3O_4 nanocubes (which is a reactive electrocatalyst) coexist.

In the modified alloy composition ($Ti_{1.4}V_{0.6}Ni$ @ C @ Co_3O_4), a selected hydrogen storage alloy, i.e., $Ti_{1.4}V_{0.6}Ni$ is covered by an electrochemical reactive Co_3O_4 nanostructures utilizing zeolite framework (Lin et al. 2016). Here, the high conductivity of carbon (C) is combined with catalytic property of nanostructured Co_3O_4. As a result, an enhanced electrochemical activity, stability, and hydrogen storage performance of the nanocomposites are achieved, much higher than the alloy itself, for example, a capacity conservation rate of 56.1% is obtained, which is remarkably larger than $Ti_{1.4}V_{0.6}Ni$ with a capacity conversion rate of 25.3%. On the other hand, a study of the kinetic properties of the above nanocomposites demonstrates an accelerating charge-transfer reaction as compared to alloy.

In summary, M-Co-Os can be favorably used as a potential material for the hydrogen storage system, not only due to their electrochemical reactivates, but also their redox properties of Co. The controlled assembly of Co-Os and M-Co-Os in combination with reactive substrates can generate remarkable composite materials for a typical hydrogen storage system.

5.6.5 Manganese (^{25}Mn)

Manganese is a known transition metal in various catalysts. The synergistic effect of Mn-O in MMOs structure is reported as a new route to design advanced catalysts for energy storage/conversion (Tan et al. 2012). Various manganese oxides (MnO$_x$), either crystalline or amorphous, including MnO_2, MnO, Mn_2O_3, Mn_3O_4, and Mn_5O_8 have been reported as catalysts due to their low cost and environmental concerns (Moreo et al. 1999). In addition, the magnetic, electronic, and electrochemical reactivity of Mn containing MMOs has achieved great interest in various fields of sustainable energy and conversions (Ma et al. 2008; Moritomo et al. 1995).

Crednerit $CuMnO_2$ ceramics is used as a hydrogen storage MMO (Abdel-Hameed et al. 2014). The average gravimetric capacity of this material is obtained to be around 16 g H$_2$/kg at 473 and about 50 g H$_2$/kg at 573 K. It is expressed that the hydrogen adsorption on the surface of $CuMnO_2$ is stable, therefore, no hydrogen evolved during desorption cycle. In addition, the reduction of Cu ions into Cu metal occurs, which may explain the irreversibility of the hydrogen adsorption process.

Li-battery material, i.e., $Li_2CoMn_3O_8$ nanoparticles are examined in an alkaline hydrogen storage system where a high discharge capacity of around 2248 mAh/g is achieved (Ghiyasiyan-Arani and Salavati-Niasari 2018). In this study, a clay template is used in order to stabilize the hydrogen storage capacity by improving structural properties like pore size and pore size distribution. A pore size of 2 nm is reported for $Li_2CoMn_3O_8$/K10 nanocomposites. In this composite system, the multisite hydrogen sorption exists either by surface adsorption of MMO nanoparticles or by penetration through the silicate layers of clay. It seems that the complete interactions by intercalation or exfoliation can positively affect the hydrogen storage performance of the product.

5.7 Summary

The basic advantage of physical adsorption, as compared to chemical absorption, is reversibility, thermodynamic preference, and fast kinetics, however, it has very low adsorption enthalpy, which can depress high storage capacity at very low temperatures. Surface modification, chemical modification, nanodispersion are effective ways for improving the thermodynamics and kinetics. Designing of novel materials for practical applications in hydrogen storage systems like fuel cells are urgently required. There are still serious challenges in this field, especially in terms of physical-chemistry of the host materials. In addition to all the above interferences, the host materials in the hydrogen storage system MUST include improved volumetric and gravimetric hydrogen densities, appropriate thermodynamics, rapid kinetics, and low cost.

Acknowledgements Universiti Sains Malaysia research grant (1001/PTEKIND/8014124), and Universiti Sains Malaysia Postdoctoral Scheme.

References

S.A.M. Abdel-Hameed, F.H. Margha, A.A. El-Meligi, Investigating hydrogen storage behavior of $CuMnO_2$ glass-ceramic material. Int. J. Energy Res. **38**, 459–465 (2014). https://doi.org/10.1002/er.3102

F.K. Butt, C. Cao, Q. Wan, P. Li, F. Idrees, M. Tahir, W.S. Khan, Z. Ali, M.J.M. Zapata, M. Safdar, X. Qu, Synthesis, evolution and hydrogen storage properties of ZnV_2O_4 glomerulus nano/microspheres: a prospective material for energy storage. Int. J. Hydrogen Energy **39**, 7842–7851 (2014). https://doi.org/10.1016/j.ijhydene.2014.03.033

R.E. Cohen, I.I. Mazin, D.G. Isaak, Magnetic collapse in transition metal oxides at high pressure: implications for the earth. Science (80-.) **275**, 654–657 (1997). https://doi.org/10.1126/science.275.5300.654

G. Deng, Y. Chen, M. Tao, C. Wu, X. Shen, H. Yang, M. Liu, Preparation and electrochemical properties of $La_{0.4}Sr_{0.6}FeO_3$ as negative electrode of Ni/MH batteries. Int. J. Hydrogen Energy **34**, 5568–5573 (2009). https://doi.org/10.1016/j.ijhydene.2009.04.061

G. Deng, Y. Chen, M. Tao, C. Wu, X. Shen, H. Yang, M. Liu, Electrochemical properties and hydrogen storage mechanism of perovskite-type oxide $LaFeO_3$ as a negative electrode for Ni/MH batteries. Electrochim. Acta **55**, 1120–1124 (2010). https://doi.org/10.1016/j.electacta.2009.09.078

T. Esaka, H. Sakaguchi, S. Kobayashi, Hydrogen storage in proton-conductive perovskite-type oxides and their application to nickel–hydrogen batteries. Solid State Ionics **166**, 351–357 (2004). https://doi.org/10.1016/j.ssi.2003.11.023

M. Ghiyasiyan-Arani, M. Salavati-Niasari, Effect of $Li_2CoMn_3O_8$ nanostructures synthesized by a combustion method on montmorillonite K10 as a potential hydrogen storage material. J. Phys. Chem. C **122**, 16498–16509 (2018). https://doi.org/10.1021/acs.jpcc.8b02617

T. Gholami, M. Salavati-Niasari, Effects of copper: aluminum ratio in CuO/Al_2O_3 nanocomposite: electrochemical hydrogen storage capacity, band gap and morphology. Int. J. Hydrogen Energy **41**, 15141–15148 (2016). https://doi.org/10.1016/j.ijhydene.2016.06.191

T. Gholami, M. Salavati-Niasari, S. Varshoy, Investigation of the electrochemical hydrogen storage and photocatalytic properties of $CoAl_2O_4$ pigment: green synthesis and characterization. Int. J. Hydrogen Energy **41**, 9418–9426 (2016). https://doi.org/10.1016/j.ijhydene.2016.03.144

T. Gholami, M. Salavati-Niasari, M. Sabet, A. Abbasi, Comparison of electrochemical hydrogen storage and Coulombic efficiency of $ZnAl_2O_4$ and $ZnAl_2O_4$-impregnated TiO_2 synthesized using green method. J. Clean. Prod. **180**, 587–594 (2018a). https://doi.org/10.1016/j.jclepro.2018.01.195

T. Gholami, M. Salavati-Niasari, A. Salehabadi, M. Amiri, M. Shabani-Nooshabadi, M. Rezaie, Electrochemical hydrogen storage properties of $NiAl_2O_4/NiO$ nanostructures using TiO_2, SiO_2 and graphene by auto-combustion method using green tea extract. Renew. Energy **115**, 199–207 (2018b). https://doi.org/10.1016/j.renene.2017.08.037

J. Haber, A. Kozlowska, R. Kozłowski, The structure and redox properties of vanadium oxide surface compounds. J. Catal. **102**, 52–63 (1986). https://doi.org/10.1016/0021-9517(86)90140-5

G. Huo, X. Lu, G. Liang, Mechanism of electrochemical hydrogen storage for α-Fe$_2$O$_3$ particles as anode material for aqueous rechargeable batteries. J. Electrochem. Soc. **163**, H566–H569 (2016). https://doi.org/10.1149/2.1011607jes

H.T. Hwang, A. Varma, Hydrogen storage for fuel cell vehicles. Curr. Opin. Chem. Eng. (2014). https://doi.org/10.1016/j.coche.2014.04.004

M. Ismail, N.S. Mustafa, N.A. Ali, N.A. Sazelee, M.S. Yahya, The hydrogen storage properties and catalytic mechanism of the CuFe$_2$O$_4$-doped MgH$_2$ composite system. Int. J. Hydrogen Energy **44**, 318–324 (2018). https://doi.org/10.1016/j.ijhydene.2018.04.191

Q. Li, N.J. Bjerrum, Aluminum as anode for energy storage and conversion: a review. J. Power Sources **110**, 1–10 (2002). https://doi.org/10.1016/S0378-7753(01)01014-X

M.M. Li, Y. Wang, C.C. Yang, Q. Jiang, In situ grown Co$_3$O$_4$ nanocubes on N-doped graphene as a synergistic hybrid for applications in nickel metal hydride batteries. Int. J. Hydrogen Energy **43**, 18421–18435 (2018). https://doi.org/10.1016/j.ijhydene.2018.08.054

J. Lin, L. Sun, Z. Cao, D. Yin, F. Liang, Y. Wu, L. Wang, A novel method to prepare Ti$_{1.4}$V$_{0.6}$Ni alloy covered with carbon and nanostructured Co$_3$O$_4$, and its good electrochemical hydrogen storage properties as negative electrode material for Ni-MH battery. Electrochim. Acta **222**, 1716–1723 (2016). https://doi.org/10.1016/j.electacta.2016.11.163

S.-B. Ma, K.-W. Nam, W.-S. Yoon, X.-Q. Yang, K.-Y. Ahn, K.-H. Oh, K.-B. Kim, Electrochemical properties of manganese oxide coated onto carbon nanotubes for energy-storage applications. J. Power Sources **178**, 483–489 (2008). https://doi.org/10.1016/J.JPOWSOUR.2007.12.027

J.B. MacChesney, H.J. Guggenheim, Growth and electrical properties of vanadium dioxide single crystals containing selected impurity ions. J. Phys. Chem. Solids **30**, 225–234 (1969). https://doi.org/10.1016/0022-3697(69)90303-5

A. Moreo, S. Yunoki, E. Dagotto, Y. Moritomo, Y. Tokura, Phase separation scenario for manganese oxides and related materials. Science **283**, 2034–2040 (1999). https://doi.org/10.1126/SCIENCE.283.5410.2034

Y. Moritomo, Y. Tomioka, A. Asamitsu, Y. Tokura, Y. Matsui, Magnetic and electronic properties in hole-doped manganese oxides with layered structures: La$_{1-x}$Sr$_{1+x}$MnO$_4$. Phys. Rev. B **51**, 3297–3300 (1995). https://doi.org/10.1103/PhysRevB.51.3297

N.S. Mustafa, N.N. Sulaiman, M. Ismail, Effect of SrFe$_{12}$O$_{19}$ nanopowder on the hydrogen sorption properties of MgH$_2$. RSC Adv. **6**, 110004–110010 (2016). https://doi.org/10.1039/c6ra22291a

D.S. Pyle, E.M. Gray, C.J. Webb, Hydrogen storage in carbon nanostructures via spillover. Int. J. Hydrogen Energy **41**, 19098–19113 (2016). https://doi.org/10.1016/J.IJHYDENE.2016.08.061

F.S. Razavi, M.S. Morassaei, A. Salehabadi, M. Ghiyasiyan-Arani, M. Salavati-Niasari, Structural characterization and electrochemical hydrogen sorption performances of the polycrystalline Ba$_2$Co$_9$O$_{14}$ nanostructures. J. Alloys Compd. **777**, 252–258 (2019). https://doi.org/10.1016/J.JALLCOM.2018.11.019

D. Saha, Z. Wei, S. Deng, Equilibrium, kinetics and enthalpy of hydrogen adsorption in MOF-177. Int. J. Hydrogen Energy **33**, 7479–7488 (2008). https://doi.org/10.1016/j.ijhydene.2008.09.053

M.Y. Saïdi, J. Barker, H. Huang, J.L. Swoyer, G. Adamson, Electrochemical properties of lithium vanadium phosphate as a cathode material for lithium-ion batteries. Electrochem. Solid-State Lett. **5**, A149 (2002). https://doi.org/10.1149/1.1479295

A. Salehabadi, M. Salavati-Niasari, Self-assembled Sr$_3$Al$_2$O$_6$-CuPc nanocomposites: a potential electrochemical hydrogen storage material. Int. J. Mater. Sci. Eng. **6**, 10–17 (2018). https://doi.org/10.17706/ijmse.2018.6.1.10-17

A. Salehabadi, M. Salavati-Niasari, T. Gholami, Effect of copper phthalocyanine (CuPc) on electrochemical hydrogen storage capacity of BaAl$_2$O$_4$/BaCO$_3$ nanoparticles. Int. J. Hydrogen Energy **42**, 15308–15318 (2017a). https://doi.org/10.1016/J.IJHYDENE.2017.05.028

A. Salehabadi, M. Salavati-Niasari, F. Sarrami, A. Karton, Sol-gel auto-combustion synthesis and physicochemical properties of BaAl$_2$O$_4$ nanoparticles; electrochemical hydrogen storage performance and density functional theory. Renew. Energy **114**, 1419–1426 (2017b). https://doi.org/10.1016/j.renene.2017.07.119

Ali Salehabadi, M. Salavati-Niasari, M. Ghiyasiyan-Arani, Self-assembly of hydrogen storage materials based multi-walled carbon nanotubes (MWCNTs) and $Dy_3Fe_5O_{12}$ (DFO) nanoparticles. J. Alloys Compd. **745**, 789–797 (2018a). https://doi.org/10.1016/J.JALLCOM.2018.02.242

Ali Salehabadi, M. Salavati-Niasari, T. Gholami, Green and facial combustion synthesis of $Sr_3Al_2O_6$ nanostructures; a potential electrochemical hydrogen storage material. J. Clean. Prod. **171**, 1–9 (2018b). https://doi.org/10.1016/J.JCLEPRO.2017.09.250

A. Salehabadi, M. Salavati-Niasari, T. Gholami, A. Khoobi, $Dy_3Fe_5O_{12}$ and $DyFeO_3$ nanostructures: green and facial auto-combustion synthesis, characterization and comparative study on electrochemical hydrogen storage. Int. J. Hydrogen Energy **43**, 9713–9721 (2018c). https://doi.org/10.1016/J.IJHYDENE.2018.04.018

A. Salehabadi, F. Sarrami, M. Salavati-Niasari, T. Gholami, D. Spagnoli, A. Karton, $Dy_3Al_2(AlO_4)_3$ ceramic nanogarnets: sol-gel auto-combustion synthesis, characterization and joint experimental and computational structural analysis for electrochemical hydrogen storage performances. J. Alloys Compd. **744**, 574–582 (2018d). https://doi.org/10.1016/j.jallcom.2018.02.117

A. Salehabadi, M.I. Ahmad, N. Morad, M. Salavati-Niasari, M. Enhessari, Electrochemical hydrogen storage properties of $Ce_{0.75}Zr_{0.25}O_2$ nanopowders synthesized by sol-gel method. J. Alloys Compd. **790**, 884–890 (2019). https://doi.org/10.1016/J.JALLCOM.2019.03.160

A. Salehabadi, N. Morad, M.I. Ahmad, A study on electrochemical hydrogen storage performance of β-copper phthalocyanine rectangular nanocuboids. Renew. Energy **146**, 497–503 (2020). https://doi.org/10.1016/J.RENENE.2019.06.176

B. Sambandam, V. Soundharrajan, Sungjin Kim, M.H. Alfaruqi, J. Jo, Seokhun Kim, V. Mathew, Y. Sun, J. Kim, Aqueous rechargeable Zn-ion batteries: an imperishable and high-energy $Zn_2V_2O_7$ nanowire cathode through intercalation regulation. J. Mater. Chem. A **6**, 3850–3856 (2018). https://doi.org/10.1039/C7TA11237H

L. Song, S. Wang, C. Jiao, X. Si, Z. Li, Shuang Liu, Shusheng Liu, C. Jiang, F. Li, J. Zhang, L. Sun, F. Xu, F. Huang, Thermodynamics study of hydrogen storage materials. J. Chem. Thermodyn. **46**, 86–93 (2012). https://doi.org/10.1016/j.jct.2011.06.022

R. Strobel, J. Garche, P.T. Moseley, L. Jorissen, G. Wolf, Advances in hydrogen storage in carbon materials. J. Power Sources **159**, 781–801 (2006). https://doi.org/10.1016/j.jpowsour.2006.03.047

Y. Tan, C. Xu, G. Chen, X. Fang, N. Zheng, Q. Xie, Facile synthesis of manganese-oxide-containing mesoporous nitrogen-doped carbon for efficient oxygen reduction. Adv. Funct. Mater. **22**, 4584–4591 (2012). https://doi.org/10.1002/adfm.201201244

K.M. Thomas, Hydrogen adsorption and storage on porous materials. Catal. Today **120**, 389–398 (2007). https://doi.org/10.1016/j.cattod.2006.09.015

B. Viswanathan, Chapter 10—Hydrogen storage, in *Energy Sources* (2017), pp. 185–212. https://doi.org/10.1016/B978-0-444-56353-8.00010-1

H.W. Verleur, A.S. Barker, C.N. Berglund, Optical properties of VO_2 between 0.25 and 5 eV. Phys. Rev. **172**, 788–798 (1968). https://doi.org/10.1103/PhysRev.172.788

Q. Wan, P. Li, J. Shan, F. Zhai, Z. Li, X. Qu, Superior catalytic effect of nickel ferrite nanoparticles in improving hydrogen storage properties of MgH_2. J. Phys. Chem. C (2015). https://doi.org/10.1021/jp508528k

J. Wang, A.D. Ebner, J.A. Ritter, On the reversibility of hydrogen storage in novel complex hydrides, in *Adsorption* (2005), pp. 811–816. https://doi.org/10.1007/s10450-005-6028-y

H. Wu, J. Du, F. Cai, F. Xu, W. Wei, J. Guo, Z. Lan, Catalytic effects of V and V_2O_5 on hydrogen storage property of $Mg_{17}Al_{12}$ alloy. Int. J. Hydrogen Energy **43**, 14578–14583 (2018). https://doi.org/10.1016/j.ijhydene.2018.06.066

C. Yuan, H.Bin Wu, Y. Xie, X.W.D. Lou, Mixed transition-metal oxides: design, synthesis, and energy-related applications. Angew. Chem. Int. Ed. **53**, 1488–1504 (2014). https://doi.org/10.1002/anie.201303971

F. Zhai, P. Li, A. Sun, S. Wu, Q. Wan, W. Zhang, Y. Li, L. Cui, X. Qu, Significantly improved dehydrogenation of $LiAlH_4$ destabilized by $MnFe_2O_4$ nanoparticles. J. Phys. Chem. C **116**, 11939–11945 (2012). https://doi.org/10.1021/jp302721w

Chapter 6
Boosting Hydrogen Storage Performances of Solid-State Materials

Abstract The term "energetic materials" are a class of material, which can release stored molecular chemical energy via external stimulations or internal modifications. We aim to take advantage of these opportunities by bringing some logical ideas onto/into the surface of hydrogen storage materials. In addition, hydrogen energy storage systems provide multiple opportunities to enhance flexibility and improve the economics of energy supply systems in the electric grid, gas pipeline systems, and transportation fuels; therefore, it is critical to boost hydrogen storage performance of the materials. The high mobility of the hydrogen and their variable compositions can be enhanced by improving the properties of the host media. In this chapter, the most important factors, which can affect the hydrogen storage performance of the solid-state materials, will be discussed.

6.1 Introduction

Hydrogen can be stored in materials either chemically (chemisorption) or physically (physisorption) under diverse conditions. In chemisorption process, the hydrogen is stored via a chemical reaction (Williamson et al. 2004). The materials in this class are limited to ammonia (NH_3), metal hydrides, carbohydrates, synthetic hydrocarbons, and liquid organic hydrogen carriers (LOHC).

Physisorption is a process in which the hydrogen molecules are weakly adsorbed at the surface of the materials. The hydrogen physisorption is kinetically favorable due to the possibility of maintaining the molecular identity of hydrogen. As mentioned before, according to the US-DOE, the most widely studied materials are porous materials, such as carbon materials (fullerenes, Nanotubes and grapheme) (Figueiredo 2018), zeolites (Weitkamp 2009), metal–organic frameworks (MOFs) (Salehabadi et al. 2020), covalent organic frameworks (COFs) (Xia and Liu 2016), microporous metal coordination materials (MMOMs) (Ozturk et al. 2016), clathrates and organotransition metal complexes (Ozturk et al. 2016).

One of the main deficiencies of the materials in the solid-state hydrogen storage technologies is significantly low specific capacity. Therefore, various experimental actions are required in order to overcome this drawback. Physisorption of the hydrogen onto the surface of the materials, normally with high surface area, offers high hydrogen storage capacities. An active and weak van der Waals forces between the hydrogen molecules and the substrates (adsorbents), governs physical adsorption reaction. A few factors need to be considered in order to enhance the hydrogen storage properties of the materials. The most important factors for boosting hydrogen storage materials will be discussed in this section.

6.2 Specific Surface Area

Specific surface area (SSA) is a property of solid materials. It is defined as the total surface area per unit of mass. In other word, SSA is a portion of the total surface area that is available for adsorption. High SSA provides reactive sites for enhancing the adsorption of the target (Germain et al. 2007). The specific surface area of a particle is a function of porosity, pore size distribution, shape, size, and roughness. The role of the specific surface area is critical in the design of a heterogeneous catalyst where typically a domain with high specific surface area (e.g., γ-alumina, silica, zeolites) denominated carrier is included in the structure of the catalyst. Polymers, MOFs, and carbon materials are three known substrates for hydrogen sorption, bearing a very high SSA. For the materials with small to medium surface area like MMOs, it is critical to enhance the surface area, in order to improve their catalytic and hydrogen storage properties. This can obtain by nanoscale formation, change in surface morphology, surface functionalization, the addition of redox species, etc.

6.2.1 Nanoscale Formation

Decreasing the particles dimension to some nanometers is essential for increasing SSA. It is logical that as the surface area per mass of a material increases, a greater amount of the particles can come into contact. Nanomaterials with high SSA can also be used favorably as an active host for hydrogen sorption. These materials show significant activity toward hydrogen sorption, however, the extent of agglomeration in nanomaterials are high, and the particles form slit-shaped pores which consist of nonuniform sizes and/or shapes (Razavi et al. 2019). For example, in a study of hydrogen storage performances of $Dy_3Fe_5O_{12}$ and $DyFeO_3$ nanoparticles, the particle sizes of $DyFeO_3$ (16–18 nm) is reported to be smaller than $Dy_3Fe_5O_{12}$ (25–30 nm) (Salehabadi et al. 2018). This dimension is further directly reflected in the hydrogen contents, where the discharge capacity of $DyFeO_3$ is higher than $Dy_3Fe_5O_{12}$.

6.2.2 *Morphology*

The synergistic effect between morphology and surface area of the materials for their catalysis effects has been reported (Cheng et al. 2014). Variation in surface morphology of the particles can affect the surface area in the following criteria:

Surface roughness and skewness—roughness and skewness are two important components of surface texture. Surface roughness is defined as the deviations in the direction from the ideal form while skewness is an asymmetry in a statistical distribution. At a microscopic level, smoothing of the surface roughness decreases the surface area. For example, it is reported that the amount of surface hydroxyl groups in $LiNbO_3$, in a study on the surface morphology and surface area on hydrogen isotopes release, can be decreased by smoothing the surface roughness through a thermal treatment (Zhu et al. 2014).

In order to calculate the surface roughness factor of the materials, the real surface area of a particle and the surface area of a sphere of the same diameter are required. The surface area of the particles can be obtained from BET results. The roughness value (R) can be calculated from BET surface area (S), density of solid (d), and its respective average particle diameter (D) (which can measure from microscopic analyses) using Eq. 6.1 (Yekeler and Ulusoy 2004);

$$R = S \times d \times (D/6) \tag{6.1}$$

For an instant, the surface roughness of $Ag_xCu_yNi_z$ alloy is reported by Rajagopal et al. (2018). They synthesized a series of alloys with different compositions of Ag, Cu, and Ni. They reported the highest value of average surface roughness for $Ag_{46}Cu_{40}Ni_{14}$ of 4.23 nm and a high skewness value of -1.66, and deduced that this sample contained a large amount of porosity. In this study, this sample is selected as an optimal sample for catalytic application. The calculated results are confirmed by AFM analysis.

Porosity—A feature of porous materials is their large value of SSA. It is known that the low packing densities associated with the conventional porous materials (carbon materials, zeolites, MOFs, etc.) can depress the volumetric capacity of hydrogen storage systems. Therefore, efforts have been devoted to overcome these deficiencies using MMOs at more moderate temperatures and pressures (Ren and North 2014). In the porous materials, the available surface area for direct contact with the surrounding environment is higher than nonporous materials (Pahalagedara et al. 2014). The presence of porosity can produce local and timely sites for hydrogen sorption (Tan et al. 2012). Hollow nanostructures are a class of porous materials with approximately high surface area, which can prepare via (a) preparation using templates and (b) preparation without templates (Arafat et al. 2014).

Shape—It is found that the properties of the materials are not only related to their sizes but also to their shapes (Xiao et al. 2011). High SSA can be obtained by controlling the shapes of the nanomaterials. The common types of crystalline

MMOs, on the basis of their morphology, are nanoparticles, nanorods, nanotubes, nanolayers, etc., which have a significant aspect ratio (Froudakis 2011).

Agglomeration—Agglomeration is an important deficiency of many nanoparticles, which can depress SSA factor. It is known that the agglomeration can form slit-shaped pores consisting of nonuniform sizes and shapes, which would further affect hydrogen content (Razavi et al. 2019). Therefore, it is critical to control the extent of agglomeration, in order to keep the nature of the particle properties. Conventionally, agglomeration of colloids was controlled by electrostatic or steric stabilization. The former was processed by increasing the electrostatic repulsion between particles, while the latter is controlled either chemically or physically (surfactants or polymers), to the surfaces (Dylla-Spears et al. 2013).

6.3   Surface Modification/Functionalization

Surface properties of modern materials are usually inadequate in terms of wettability, adhesion properties, biocompatibility, etc.; therefore, they must be modified prior to application (Mozetič 2019). In general, surface modification is defined as any action, which can change the physical, chemical, or biological characteristics of the materials as compared to ones originally found on the surface of a material (Fig. 6.1). For example, the hydrophilic nature of hydroxyl groups on MMOs can form micron-sized agglomerates (Pourrahimi 2016). Generally, the alkyls containing amine (R-NH$_2$ and R-NH-R′), hydroxyl (R-OH), and carboxyl (R-COOH) functional groups are appropriate for surface modification. In addition, the polymers and silicone templates are used for surface modification of MMOs, in order to enlarge their SSA and pore size (Yu Lu et al. 2002).

In a new approach, mesostructured crystalline bimetal oxides are formed in one single impregnation step under reflux using template. Under these conditions, NiFe$_2$O$_4$ (250 m^2 g^{-1}), and CuFe$_2$O$_4$ (296 m^2 g^{-1}) are prepared in a very high SSA (Yen et al. 2011).

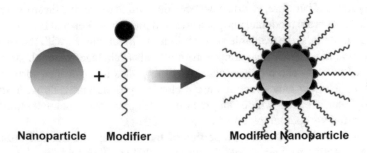

Nanoparticle   Modifier   Modified Nanoparticle

Fig. 6.1 Surface modification of nanoparticles

The modification of the nanoparticles is also reported, utilizing sensitizers such as copper phthalocyanine (CuPc), which bonded on the surface of the nanoparticles. Experimentally, in a study on electrochemical hydrogen storage performances of $BaAl_2O_4$ and $Sr_3Al_6O_{12}$ nanoparticles, it is observed that upon surface modification of the particles, their discharge capacities, and their respective hydrogen contents are enhanced (Salehabadi et al. 2017a; Salehabadi and Salavati-Niasari 2018).

Incorporation of functional groups is a renowned surface modification technique that includes methods such as oxidation, amination, nitration, and halogenation. Lack of stability of the treated surface is a severe problem. Radiation and plasma treatments are two techniques of this class of surface modification.

6.4 Redox Species

In the reaction of inorganic compounds with active organic compounds, generally, the electrons transfer from one to another. According to the basic chemistry, electron gain/loss is called reduction/oxidation, where the joint process is called a "redox reaction". Redox-reactions can release energy and in combustion or energy storage technologies (Zeng et al. 2018). The mechanism of electrochemical hydrogen storage can be considered to be probably due to the valence change usually in the ions (for example A and/or B in an MMO with a general formula of $A_xB_yO_z$) (Esaka et al. 2004). In proton conductor MMOs like perovskites containing zirconium (Zr) and/or cerium (Ce), it was observed that in the presence of the only Zr^{4+}, the discharge property fell off, while in Ce^{4+} containing perovskite, the hydrogen discharge performance is gradually enhanced. This phenomenon is due to the reduction effect of Ce ($Ce^{4+} \rightarrow Ce^{3+}$), which is preferable to Zr^{4+} (Iwahara et al. 1993). Hence, irrespective of the general aspect, it should be pointed out that in the MMOs, the central ions are not the only species that can affect the hydrogen storage. Therefore, the process of hydrogen sorption for a typical MMOs containing binary redox species would be followed as Eq. 6.2;

$$A_x^n B_y^m O_z + f H_2O + f e^- \leftrightarrow A_{x-x'}^n A_x^{n-1} B_{y-y'}^m B_y^{m-1} O_z - H_f + f OH^-, \quad (6.2)$$

where "n" and "m" are oxidation states of the ions, and "f" is the reaction mass balance.

The mechanism of electrochemical hydrogen storage α-Fe_2O_3 is also reported by Huo et al. (2016). They proposed that the α-Fe_2O_3 can be potentially utilized as an anode material for aqueous rechargeable batteries. For the sequences of charge and discharge reactions of α-Fe_2O_3, they theoretically assumed two conditions;

(a) the reaction of Fe^{3+} and OH^- occurs only on the surface, and
(b) H^+ is able to enter the crystal lattice and proposed a series of reactions for charging process Eq. 6.3;

$$Fe_2O_3 + 2H^+ + 2e^- \rightarrow Fe(OH)_2 + FeO,$$
$$\text{and} \quad FeO + 2H^+ + 2e^- \rightarrow Fe + H_2O \tag{6.3}$$

and discharging process Eq. 6.4;

$$Fe + 2H_2O - 2e^- \rightarrow Fe(OH)_2 + 2H^+,$$
$$\text{and} \quad 3Fe(OH)_2 - 2e^- \rightarrow Fe_3O_4 + 2H_2O + 2H^+. \tag{6.4}$$

In the charging process (Eq. 6.3), Fe^{3+} is reduced to Fe^{2+}, where Fe^{2+} is oxidized during discharge (Eq. 6.4).

In other studies, $LaFeO_3$ (Deng et al. 2010a) and $LaCrO_3$ (Deng et al. 2010b) are reported as negative electrodes for Ni/MH batteries and their hydrogen storage performances are discussed. In these researches, redox effects of Fe ($Fe^{3+} \rightarrow Fe^{2+}$) and Cr ($Cr^{3+} \rightarrow Cr^{2+}$) are proposed for superior hydrogen storage performances. They concluded that these electrodes are promising for application in secondary batteries, as no activation is required for charging and discharging processes. The mechanism of hydrogen adsorption is also predicted by Salehabadi et al. (2018) in a study on $Dy_3Fe_5O_{12}$ nanostructures. In this research, in addition to the general hydrogen physisorption reaction, a complementary reaction mechanism based on Fe^{3+} reduction is proposed as Eq. 6.5;

$$Dy_x Fe_y^{(III)} O_z + x H_2O + xe^- \leftrightarrow Dy_x Fe_{y-n}^{(III)} Fe_n^{(II)} O_z - H_x + x OH^-. \tag{6.5}$$

It can be concluded that the presence of redox species in the structure of MMOs can potentially enhance the hydrogen storage performances, as compared to the sample without valence change.

6.5   Electronically Active Center

Energetic materials are a class of material that can release chemical energy stored in their molecular structure. The presence of metallic conductors and semiconductors allow more successful interaction of hydrogen with the electronic structure of the host (Seenithurai and Chai 2018). Theoretically, it is known that the crystal structure and its electronic structure of the host center (metal, semiconductors) can be changed upon the phase transition (Züttel 2003). For example, a hydrogen storage MMOs containing Dy and Fe ions, has a discharge capacity of around 2000 mAh g^{-1} (Salehabadi et al. 2018), while in the spinel MMOs with Ba and Al ions, the discharge capacity is found to be around 1000 mAh g^{-1} (Salehabadi et al. 2017b).

6.6 Temperature

In hydrogen storage materials, the behavior of the hosts can be described by a pressure-composition-temperature (PCT) curve (Suib et al. 2013b). This curve is a plot of pressure versus composition at various temperatures. It is found that the hydrogen desorption temperature and reaction rates are higher at elevated temperatures (Suib et al. 2013a). In an electrochemical hydrogen storage $LaCrO_3$, Deng and his coworkers (Deng et al. 2010b) computed the effect of temperature on the hydrogen storage performances of the host materials in a setup of the testing instrument at 298, 313 and 333 K in 7 M KOH solution. They reported that the initial discharge capacity of the working electrodes is up to 484.1 mAh g^{-1} at 333 K, much higher than that at 298 K (194.8 mAh g^{-1}). Therefore, owing to the thermal stability of the MMOs, they can be preferably used as a host for harvesting hydrogen storage materials.

6.7 Summary

The hydrogen storage and delivery are two pertinent technical issues that require materials improvement before the potential of hydrogen storage is established. The most relevant and recent approaches toward hydrogen storage are solid-state materials with attention to their structures and properties. The superiority of solid-state materials is highlighted based on their stability and efficiency. In addition to that, various criteria should be considered in determining an efficient hydrogen storage system. The criteria are selection of at least one active central ion with wide range oxidation states (redox species), surface modification, fabrication of composite/hybrid/doped materials, and fabrication of uneven surface morphology with large surface area and pore size.

Acknowledgements Universiti Sains Malaysia research grant (1001/PTEKIND/8014124), and Universiti Sains Malaysia Postdoctoral Scheme.

References

M.M. Arafat, A.S.M.A. Haseeb, S.A. Akbar, Developments in semiconducting oxide-based gas-sensing materials, in *Comprehensive Materials Processing* (Elsevier Ltd, Amsterdam, 2014), pp. 205–219. https://doi.org/10.1016/B978-0-08-096532-1.01307-8

H. Cheng, J. Wang, Y. Zhao, X. Han, Effect of phase composition, morphology, and specific surface area on the photocatalytic activity of TiO_2 nanomaterials. RSC Adv. **4**, 47031–47038 (2014). https://doi.org/10.1039/c4ra05509h

G. Deng, Y. Chen, M. Tao, C. Wu, X. Shen, H. Yang, M. Liu, Electrochemical properties and hydrogen storage mechanism of perovskite-type oxide $LaFeO_3$ as a negative electrode for Ni/MH batteries. Electrochim. Acta **55**, 1120–1124 (2010a). https://doi.org/10.1016/j.electacta.2009.09.078

G. Deng, Y. Chen, M. Tao, C. Wu, X. Shen, H. Yang, M. Liu, Study of the electrochemical hydrogen storage properties of the proton-conductive perovskite-type oxide $LaCrO_3$ as negative electrode for Ni/MH batteries. Electrochim. Acta **55**, 884–886 (2010b). https://doi.org/10.1016/j.electacta.2009.06.071

R. Dylla-Spears, M.D.L. Feit, P.E. Miller, W.A. Steele, I.T. Suratwala, L.L. Wong, Method for preventing agglomeration of charged colloids without loss of surface activity. WO2014070461A1, 2013

T. Esaka, H. Sakaguchi, S. Kobayashi, Hydrogen storage in proton-conductive perovskite-type oxides and their application to nickel–hydrogen batteries. Solid State Ionics **166**, 351–357 (2004). https://doi.org/10.1016/J.SSI.2003.11.023

J.L. Figueiredo, Nanostructured porous carbons for electrochemical energy conversion and storage. Surf. Coat. Technol. **350**, 307–312 (2018). https://doi.org/10.1016/J.SURFCOAT.2018.07.033

G.E. Froudakis, Hydrogen storage in nanotubes & nanostructures. Mater. Today **14**, 324–328 (2011). https://doi.org/10.1016/S1369-7021(11)70162-6

J. Germain, J.M.J. Fréchet, F. Svec, Hypercrosslinked polyanilines with nanoporous structure and high surface area: potential adsorbents for hydrogen storage. J. Mater. Chem. **17**, 4989 (2007). https://doi.org/10.1039/b711509a

G. Huo, X. Lu, G. Liang, Mechanism of electrochemical hydrogen storage for α-Fe_2O_3 particles as anode material for aqueous rechargeable batteries. J. Electrochem. Soc. **163**, H566–H569 (2016). https://doi.org/10.1149/2.1011607jes

H. Iwahara, T. Yajima, T. Hibino, K. Ozaki, H. Suzuki, Protonic conduction in calcium, strontium and barium zirconates. Solid State Ionics **61**, 65–69 (1993)

Y. Lu, Y. Yin, B.T. Mayers, Y. Xia, Modifying the surface properties of superparamagnetic iron oxide nanoparticles through a sol–gel approach. Nano Lett. **2**, 183–186 (2002). https://doi.org/10.1021/NL015681Q

M. Mozetič, Surface modification to improve properties of materials. Materials (Basel) **12**, 441 (2019). https://doi.org/10.3390/ma12030441

Z. Ozturk, D.A. Kose, Z.S. Sahin, G. Ozkan, A. Asan, Novel 2D micro-porous Metal-Organic Framework for hydrogen storage. Int. J. Hydrogen Energy **41**, 12167–12174 (2016). https://doi.org/10.1016/J.IJHYDENE.2016.05.170

M.N. Pahalagedara, L.R. Pahalagedara, C.-H. Kuo, S. Dharmarathna, S.L. Suib, Ordered mesoporous mixed metal oxides: remarkable effect of pore size on catalytic activity. Langmuir **30**, 8228–8237 (2014). https://doi.org/10.1021/la502190b

A.M. Pourrahimi, The synthesis, surface modification and use of metal-oxide nanoparticles in polyethylene for ultra-low transmission-loss HVDC cable insulation materials. KTH Royal Institute of Technology, 2016

V. Rajagopal, P. Manivel, N. Nesakumar, M. Kathiresan, D. Velayutham, V. Suryanarayanan, $Ag_xCu_yNi_z$ trimetallic alloy catalysts for the electrocatalytic reduction of benzyl bromide in the presence of carbon dioxide. ACS Omega **3**, 17125–17134 (2018). https://doi.org/10.1021/acsomega.8b02715

F.S. Razavi, M.S. Morassaei, A. Salehabadi, M. Ghiyasiyan-Arani, M. Salavati-Niasari, Structural characterization and electrochemical hydrogen sorption performances of the polycrystalline $Ba_2Co_9O_{14}$ nanostructures. J. Alloys Compd. **777**, 252–258 (2019). https://doi.org/10.1016/J.JALLCOM.2018.11.019

J. Ren, B.C. North, Shaping porous materials for hydrogen storage applications: a review. J. Technol. Innov. Renew. Energy (2014)

A. Salehabadi, N. Morad, M.I. Ahmad, A study on electrochemical hydrogen storage performance of β-copper phthalocyanine rectangular nanocuboids. Renew. Energy **146**, 497–503 (2020). https://doi.org/10.1016/J.RENENE.2019.06.176

A. Salehabadi, M. Salavati-Niasari, Self-assembled $Sr_3Al_2O_6$-CuPc nanocomposites: a potential electrochemical hydrogen storage material. Int. J. Mater. Sci. Eng. **6**, 10–17 (2018). https://doi.org/10.17706/ijmse.2018.6.1.10-17

A. Salehabadi, M. Salavati-Niasari, T. Gholami, Effect of copper phthalocyanine (CuPc) on electrochemical hydrogen storage capacity of $BaAl_2O_4$/$BaCO_3$ nanoparticles. Int. J. Hydrogen Energy **42**, 15308–15318 (2017a). https://doi.org/10.1016/J.IJHYDENE.2017.05.028

A. Salehabadi, M. Salavati-Niasari, T. Gholami, A. Khoobi, $Dy_3Fe_5O_{12}$ and $DyFeO_3$ nanostructures: green and facial auto-combustion synthesis, characterization and comparative study on electrochemical hydrogen storage. Int. J. Hydrogen Energy **43**, 9713–9721 (2018). https://doi.org/10.1016/J.IJHYDENE.2018.04.018

A. Salehabadi, M. Salavati-Niasari, F. Sarrami, A. Karton, Sol-Gel auto-combustion synthesis and physicochemical properties of $BaAl_2O_4$ nanoparticles; electrochemical hydrogen storage performance and density functional theory. Renew. Energy **114**, 1419–1426 (2017b). https://doi.org/10.1016/j.renene.2017.07.119

S. Seenithurai, J.-D. Chai, Electronic and hydrogen storage properties of Li-terminated linear boron chains studied by TAO-DFT. Sci. Rep. **8**, 13538 (2018). https://doi.org/10.1038/s41598-018-31947-9

S.L. Suib, Y. Kojima, H. Miyaoka, T. Ichikawa, Hydrogen storage materials, in *New and Future Developments in Catalysis* (Elsevier, London, 2013a), pp. 99–136. https://doi.org/10.1016/B978-0-444-53880-2.00006-5

S.L. Suib, Y. Liu, H. Pan, Hydrogen storage materials, in *New and Future Developments in Catalysis* (Elsevier, London, 2013b), pp. 377–405. https://doi.org/10.1016/B978-0-444-53880-2.00018-1

Y.H. Tan, J.A. Davis, K. Fujikawa, N.V. Ganesh, A.V. Demchenko, K.J. Stine, Surface area and pore size characteristics of nanoporous gold subjected to thermal, mechanical, or surface modification studied using gas adsorption isotherms, cyclic voltammetry, thermogravimetric analysis, and scanning electron microscopy. J. Mater. Chem. **22**, 6733–6745 (2012). https://doi.org/10.1039/C2JM16633J

J. Weitkamp, Fuels–hydrogen storage–zeolites, in *Encyclopedia of Electrochemical Power Sources*, ed. by J. Garche (Elsevier, Netherland, 2009), pp. 497–503. https://doi.org/10.1016/B978-044452745-5.00330-0

A.J. Williamson, F.A. Reboredo, G. Galli, Chemisorption on semiconductor nanocomposites: a mechanism for hydrogen storage. Appl. Phys. Lett. **85**, 2917–2919 (2004). https://doi.org/10.1063/1.1800274

L. Xia, Q. Liu, Lithium doping on covalent organic framework-320 for enhancing hydrogen storage at ambient temperature. J. Solid State Chem. **244**, 1–5 (2016). https://doi.org/10.1016/J.JSSC.2016.09.007

G. Xiao, P. Gao, L. Wang, Y. Chen, Y. Wang, G. Zhang, Ultrasonochemical-assisted synthesis of CuO nanorods with high hydrogen storage ability. J. Nanomater. **2011**, 1–6 (2011). https://doi.org/10.1155/2011/439162

M. Yekeler, U. Ulusoy, Characterisation of surface roughness and wettability of salt-type minerals: calcite and barite. Trans. Inst. Min. Metall. Sect. C: Miner. Process. Extract. Metall. **113**, 145–152. https://doi.org/10.1179/037195504225006100

H. Yen, Y. Seo, R. Guillet-Nicolas, S. Kaliaguine, F. Kleitz, One-step-impregnation hard templating synthesis of high-surface-area nanostructured mixed metal oxides ($NiFe_2O_4$, $CuFe_2O_4$ and Cu/CeO_2). Chem. Commun. **47**, 10473 (2011). https://doi.org/10.1039/c1cc13867g

L. Zeng, Z. Cheng, J.A. Fan, L.-S. Fan, J. Gong, Metal oxide redox chemistry for chemical looping processes. Nat. Rev. Chem. **2**, 349–364 (2018). https://doi.org/10.1038/s41570-018-0046-2

D. Zhu, T. Oda, S. Tanaka, Influence of surface morphology and surface area on release behavior of hydrogen isotopes in $LiNbO_3$. Fusion Eng. Des. **89**, 2797–2805 (2014). https://doi.org/10.1016/j.fusengdes.2014.08.005

A. Züttel, Materials for hydrogen storage. Mater. Today **6**, 24–33 (2003). https://doi.org/10.1016/S1369-7021(03)00922-2

Index

Printed in the United States
By Bookmasters